Improving the Sensory, Nutritional and Physicochemical Quality of Fresh Meat

Improving the Sensory, Nutritional and Physicochemical Quality of Fresh Meat

Editor

Paulo Eduardo Sichetti Munekata

MDPI • Basel • Beijing • Wuhan • Barcelona • Belgrade • Manchester • Tokyo • Cluj • Tianjin

Editor
Paulo Eduardo Sichetti Munekata
Meat Technology Center of Galicia
Spain

Editorial Office
MDPI
St. Alban-Anlage 66
4052 Basel, Switzerland

This is a reprint of articles from the Special Issue published online in the open access journal *Foods* (ISSN 2304-8158) (available at: https://www.mdpi.com/journal/foods/special_issues/sensory_nutritional_quality_fresh_meat).

For citation purposes, cite each article independently as indicated on the article page online and as indicated below:

LastName, A.A.; LastName, B.B.; LastName, C.C. Article Title. *Journal Name* **Year**, *Volume Number*, Page Range.

ISBN 978-3-0365-2287-6 (Hbk)
ISBN 978-3-0365-2288-3 (PDF)

© 2021 by the authors. Articles in this book are Open Access and distributed under the Creative Commons Attribution (CC BY) license, which allows users to download, copy and build upon published articles, as long as the author and publisher are properly credited, which ensures maximum dissemination and a wider impact of our publications.

The book as a whole is distributed by MDPI under the terms and conditions of the Creative Commons license CC BY-NC-ND.

Contents

About the Editor vii

Preface to "Improving the Sensory, Nutritional and Physicochemical Quality of Fresh Meat" . ix

Paulo E. S. Munekata
Improving the Sensory, Nutritional and Physicochemical Quality of Fresh Meat
Reprinted from: *Foods* **2021**, *10*, 2060, doi:10.3390/foods10092060 1

Noemí Echegaray, Rubén Domínguez, Vasco A. P. Cadavez, Roberto Bermúdez, Laura Purriños, Ursula Gonzales-Barron, Ettiene Hoffman and José M. Lorenzo
Influence of the Production System (Intensive vs. Extensive) at Farm Level on Proximate Composition and Volatile Compounds of Portuguese Lamb Meat
Reprinted from: *Foods* **2021**, *10*, 1450, doi:10.3390/foods10071450 3

Clément Burgeon, Alice Markey, Marc Debliquy, Driss Lahem, Justine Rodriguez, Ahmadou Ly and Marie-Laure Fauconnier
Comprehensive SPME-GC-MS Analysis of VOC Profiles Obtained Following High-Temperature Heating of Pork Back Fat with Varying Boar Taint Intensities
Reprinted from: *Foods* **2021**, *10*, 1311, doi:10.3390/foods10061311 27

Mohamed F. Eshag Osman, Abdellatif A. Mohamed, Mohammed S. Alamri, Isam Ali Mohamed Ahmed, Shahzad Hussain, Mohamed I. Ibraheem and Akram A. Qasem
Quality Characteristics of Beef Patties Prepared with Octenyl-Succinylated (Osan) Starch
Reprinted from: *Foods* **2021**, *10*, 1157, doi:10.3390/foods10061157 45

Christina Bakker, Keith Underwood, Judson Kyle Grubbs and Amanda Blair
Low-Voltage Electrical Stimulation of Beef Carcasses Slows Carcass Chilling Rate and Improves Steak Color
Reprinted from: *Foods* **2021**, *10*, 1065, doi:10.3390/foods10051065 61

Kazem Alirezalu, Milad Yaghoubi, Leila Poorsharif, Shadi Aminnia, Halil Ibrahim Kahve, Mirian Pateiro, José M. Lorenzo and Paulo E. S. Munekata
Antimicrobial Polyamide-Alginate Casing Incorporated with Nisin and ε-Polylysine Nanoparticles Combined with Plant Extract for Inactivation of Selected Bacteria in Nitrite-Free Frankfurter-Type Sausage
Reprinted from: *Foods* **2021**, *10*, 1003, doi:10.3390/foods10051003 71

Yong Geum Shin, Dhanushka Rathnayake, Hong Seok Mun, Muhammad Ammar Dilawar, Sreynak Pov and Chul Ju Yang
Sensory Attributes, Microbial Activity, Fatty Acid Composition and Meat Quality Traits of Hanwoo Cattle Fed a Diet Supplemented with Stevioside and Organic Selenium
Reprinted from: *Foods* **2021**, *10*, 129, doi:10.3390/foods10010129 85

About the Editor

Paulo Eduardo Sichetti Munekata (Ph.D.) is a postdoctoral researcher at the *Centro Tecnológico de la Carne de Galicia*, Ourense, Spain. He obtained his Ph.D. degree in Science of Food Engineering (University of São Paulo/Brazil) in 2016. His main areas of interest are centered in Food Science and Technology: Meat Science, Meat Processing, and Development of Healthier and Functional Foods. He is editorial board member in the journals *Foods, Antioxidants, and Frontiers in Animal Science*. He has published more than 200 scientific publications in peer-reviewed international journals, book chapters, and communications in national and international congresses. He also co-edited 5 books for international publishers and has h-index 29 in Scopus.

Preface to "Improving the Sensory, Nutritional and Physicochemical Quality of Fresh Meat"

Advances in the current meat chain are necessary in our society due to the increasing demand that require high quality foods. Consequently, meat producers and researchers are facing new and complex challenges to attend the demand, produce high quality meat, and develop strategies to increase its preservation. In this context, scientific advances are necessary to improve the knowledge about key factors of the meat chain that are comprised of production systems, diet composition, carcass management, volatile composition, microbial quality, and sensory analysis of fresh meat, as well as meat processing. The e-book and Special Issue titled *"Improving the Sensory, Nutritional and Physicochemical Quality of Fresh Meat"* is composed of six original papers about recent advances in the area of Meat Science about these key factors.

Paulo Eduardo Sichetti Munekata
Editor

Editorial

Improving the Sensory, Nutritional and Physicochemical Quality of Fresh Meat

Paulo E. S. Munekata

Centro Tecnológico de la Carne de Galicia, Rúa Galicia n° 4, Parque Tecnológico de Galicia, San Cibrao das Viñas, 32900 Ourense, Spain; paulosichetti@ceteca.net

Citation: Munekata, P.E.S. Improving the Sensory, Nutritional and Physicochemical Quality of Fresh Meat. *Foods* **2021**, *10*, 2060. https://doi.org/10.3390/foods10092060

Received: 24 August 2021
Accepted: 31 August 2021
Published: 1 September 2021

Publisher's Note: MDPI stays neutral with regard to jurisdictional claims in published maps and institutional affiliations.

Copyright: © 2021 by the author. Licensee MDPI, Basel, Switzerland. This article is an open access article distributed under the terms and conditions of the Creative Commons Attribution (CC BY) license (https://creativecommons.org/licenses/by/4.0/).

This Special Issue titled "Improving the Sensory, Nutritional and Physicochemical Quality of Fresh Meat" is comprised of six studies that explored different strategies to improve the quality of fresh meat, as well as some aspects related to its further processing.

The increasing demand for high quality meat has pushed the professionals and researchers of the meat production area to face new challenges. Consequently, advances covering different stages of the meat production chain process and factors have been made to increase the knowledge and develop strategies to produce and improve the preservation of high quality meat [1–3]. Therefore, studying the effect of production systems, diet composition, carcass management, volatile composition of fresh meat related to sensory properties and further processing are topics of interest in attempts to increase knowledge about meat quality and pave the way for strategic changes in the meat industry.

The meat production chain has many stages, starting with the rearing of animals using different systems and diets with optimized composition to favor the animal development and the production of meat with characteristics that are aligned with consumer preferences [3]. Regarding the influence of the production system, Echegaray et al. [4] evaluated the composition and volatile composition of lamb meat produced in an intensive or extensive system. Significant increases in intramuscular fat and protein content in *longissimus thoracis et lumborum* were obtained from animals reared in an extensive system in comparison to animals produced in an intensive system. The extensive system also led to a bigger accumulation of volatile composition in meat in relation to the intensive system. Interestingly, the authors also discussed the relation of intramuscular fat with the composition and content of lipid-derived volatile fraction, wherein these differences were attributed to the presence of specific compounds in each diet (natural antioxidants, for instance) and lipid fraction composition (unsaturated fatty acids and their susceptibility to oxidation).

Another experiment in the context of animal production was carried out by Shin et al. [5] to characterize the effect of stevioside (bioactive compound naturally found in the leaves of *Stevia rebaudiana*) and organic selenium (a crucial component for the endogenous antioxidant system in humans and animals) on the quality of Hanwoo meat. These authors observed significant improvements in animal performance (especially in weight gain and final weight) by using these supplements. The characteristics of meat obtained from animals with the supplemented diet was also improved due to the increase in protein, moisture, PUFA contents, redness, and oxidative stability, and the simultaneous reduction in total cholesterol content, shear force, and drip loss. No major effects in sensory attributes of fresh meat or microbial growth during storage were reported.

Moving forward in the meat production chain, the adequate processing of carcass is a necessary action in order to obtain high-quality meat [1]. In this sense, the experiment conducted by Bakker et al. [6] reported significant improvements in the tenderness and color of beef due to low-voltage electric stimulation of beef carcasses. Moreover, the authors of this study also reported a non-significant impact in cooking loss between the meat control and electrically stimulated carcasses.

The characteristics of meat, especially the sensory attributes, play a central role when consumers judge the acceptance of sensory attributes and the whole eating experience [7]. One of the main factors in this context is the accumulation of unpleasant volatile compounds. This context was considered in the study carried out by Burgeon et al. [8], who evaluated the formation of volatile compounds associated with the perception of boar taint (strong smell found in uncastrated male pigs associated with the presence of skatole and androstenone) in cooked pork. The authors observed higher concentrations of androstenone as the temperature was increased, which indicated a temperature dependence effect. However, the same effect was not reported for skatole. Additionally, studies with further processing of meat were also included in order to explore healthier [9] and functional [10] meat-based foods.

The advances reported in these studies can be seen as meaningful contributions to the generation of knowledge of fresh meat quality and assist, to some extent, in the strategic development of meat production considering the consumer preferences and market trends. Finally, I would like to thank the authors for their submissions; express my sincere gratitude to the reviewers for their time, effort, and insightful suggestions and comments during the reviewing process; and acknowledge the supportive, thoughtful, and expert assistance of the associated editors of Foods throughout the preparation and management of this Special Issue.

Funding: This research received no external funding.

Institutional Review Board Statement: Not applicable.

Informed Consent Statement: Not applicable.

Conflicts of Interest: The author declares no conflict of interest.

References

1. Pophiwa, P.; Webb, E.C.; Frylinck, L. A review of factors affecting goat meat quality and mitigating strategies. *Small Rumin. Res.* **2020**, *183*, 106035. [CrossRef]
2. Tomasevic, I.; Djekic, I.; Font-i-Furnols, M.; Terjung, N.; Lorenzo, J.M. Recent advances in meat color research. *Curr. Opin. Food Sci.* **2021**, *41*, 81–87. [CrossRef]
3. Borgogno, M.; Favotto, S.; Corazzin, M.; Cardello, A.V.; Piasentier, E. The role of product familiarity and consumer involvement on liking and perceptions of fresh meat. *Food Qual. Prefer.* **2015**, *44*, 139–147. [CrossRef]
4. Echegaray, N.; Domínguez, R.; Cadavez, V.A.P.; Bermúdez, R.; Purriños, L.; Gonzales-Barron, U.; Hoffman, E.; Lorenzo, J.M. Influence of the production system (Intensive vs. extensive) at farm level on proximate composition and volatile compounds of portuguese lamb meat. *Foods* **2021**, *10*, 1450. [CrossRef]
5. Shin, Y.G.; Rathnayake, D.; Mun, H.S.; Dilawar, M.A.; Pov, S.; Yang, C.J. Sensory attributes, microbial activity, fatty acid composition and meat quality traits of Hanwoo cattle fed a diet supplemented with stevioside and organic selenium. *Foods* **2021**, *10*, 129. [CrossRef] [PubMed]
6. Bakker, C.; Underwood, K.; Grubbs, J.K.; Blair, A. Low-voltage electrical stimulation of beef carcasses slows carcass chilling rate and improves steak color. *Foods* **2021**, *10*, 1065. [CrossRef] [PubMed]
7. Khan, M.I.; Jo, C.; Tariq, M.R. Meat flavor precursors and factors influencing flavor precursors—A systematic review. *Meat Sci.* **2015**, *110*, 278–284. [CrossRef] [PubMed]
8. Burgeon, C.; Markey, A.; Debliquy, M.; Lahem, D.; Rodriguez, J.; Ly, A.; Fauconnier, M.L. Comprehensive SPME-GC-MS analysis of voc profiles obtained following high-temperature heating of pork back fat with varying boar taint intensities. *Foods* **2021**, *10*, 1311. [CrossRef] [PubMed]
9. Alirezalu, K.; Yaghoubi, M.; Poorsharif, L.; Aminnia, S.; Kahve, H.I.; Pateiro, M.; Lorenzo, J.M.; Munekata, P.E.S. Antimicrobial polyamide-alginate casing incorporated with nisin and ε-polylysine nanoparticles combined with plant extract for inactivation of selected bacteria in nitrite-free frankfurter-type sausage. *Foods* **2021**, *10*, 1003. [CrossRef] [PubMed]
10. Eshag Osman, M.F.; Mohamed, A.A.; Alamri, M.S.; Ali, I.; Hussain, S.; Ibraheem, M.I.; Qasem, A.A. Quality characteristics of beef patties prepared with octenyl-succinylated (Osan) starch. *Foods* **2021**, *10*, 1157. [CrossRef] [PubMed]

Influence of the Production System (Intensive vs. Extensive) at Farm Level on Proximate Composition and Volatile Compounds of Portuguese Lamb Meat

Noemí Echegaray [1], Rubén Domínguez [1,*], Vasco A. P. Cadavez [2], Roberto Bermúdez [1], Laura Purriños [1], Ursula Gonzales-Barron [2], Ettiene Hoffman [3] and José M. Lorenzo [1,4]

[1] Centro Tecnológico de la Carne de Galicia, Avd. Galicia No 4, Parque Tecnológico de Galicia, San Cibrao das Viñas, 32900 Ourense, Spain; noemiechegaray@ceteca.net (N.E.); robertobermudez@ceteca.net (R.B.); laurapurrinos@ceteca.net (L.P.); jmlorenzo@ceteca.net (J.M.L.)

[2] Centro de Investigação de Montanha (CIMO), Instituto Politécnico de Bragança, 5300-253 Bragança, Portugal; vcadavez@ipb.pt (V.A.P.C.); ubarron@ipb.pt (U.G.-B.)

[3] Faculty of Management, Canadian University Dubai, Dubai 117781, United Arab Emirates; ettiene.hoffman@cud.ac.ae

[4] Área de Tecnología de los Alimentos, Facultad de Ciencias de Ourense, Universidad de Vigo, 32004 Ourense, Spain

* Correspondence: rubendominguez@ceteca.net

Abstract: Today's society demands healthy meat with a special emphasis on integrated animal husbandry combined with the concern for animal welfare. In this sense, the raising of lambs in an extensive system has been one of the most common practices, which results in meats with high nutritional value. However, both the production system and the diet play a fundamental role in the chemical composition of the meat, which has a direct impact on the content of volatile compounds. Thus, the aim of this study was to determine the effect of two production systems (intensive and extensive) on the chemical composition and volatile profile of lamb meat. Twenty-eight lambs of the Bordaleira-de-Entre-Douro-e-Minho (BEDM) sheep breed were raised for meat production under the intensive or extensive system and were fed with concentrate and pasture, respectively. All animals were carried out in the muscle *longissimus thoracis et lumborum*. Results evidenced that all the composition parameters were affected by the production system. Extensively-reared lambs produced meat with the highest fat and protein contents, while these animals had the lowest percentages of moisture and ash. Similarly, the total content of volatile compounds was affected ($p < 0.05$) by the production system and were higher in the meat of lambs reared extensively. Furthermore, the content of total acids, alcohols, aldehydes, esters, ethers, furans and sulfur compounds as well as most of the individual compounds were also affected ($p < 0.05$) by the production system, whereas total hydrocarbons and ketones were not affected ($p > 0.05$). As a general conclusion, the production system had very high influence not only in proximate composition but also in the volatile compounds.

Keywords: Bordaleira-de-Entre-Douro-e-Minho; rearing system; pasture; concentrate; volatile compounds

1. Introduction

The meat quality is an essential factor in ensuring consumer satisfaction [1] and is related to several parameters such as visual appearance, quality and distribution of the fat, texture, juiciness as well as flavor [2]. Specifically, in lamb meat, the odor and flavor are two of the most important eating quality attributes since the meat of these animals have a unique aroma [3–5]. In this manner, lamb meat is characterized by a typical species-related flavor that is denominated as "mutton flavor", which could seriously affect the acceptability of consumers [6,7].

On the other hand, in response to consumer demand, the sheep farming sector is increasingly concerned with incrementing the added value of its products through sustainability, animal welfare and conservation of ancient autochthonous genetic types [8,9].

In this regard, the use of autochthonous breeds for meat production is of special interest due to the promotion of the valorization, protection and conservation of the zoogenetic heritage [10]. This is the case of the Portuguese native breed, named Bordaleira-de-Entre-Douro-e-Minho (BEDM), which can also contribute to the diversity of production systems due to its particular characteristics such as local adaptation, resistance to diseases and high fertility [11,12]. These qualities allow the use of natural pastures in lamb rearing [13]. Nevertheless, the characteristics generated by extensive rearing can sometimes result in various unwanted modifications in the organoleptic quality of the lamb meat with respect to intensive commercial farming. This is the case of the volatile profile, which in addition to being influenced by the animal's genetics, slaughter age and management practices is strongly influenced by the diet supplied [6,14,15]. In fact, previous studies have linked certain volatile compounds with a specific diet [16,17]. Thus, volatile substances such as terpenoids [14,18], phenols [19] and the diketone 2,3-octanedione [14,18,20] were related with pasture-based diets; while lactones [20,21], branched fatty acids [6,20,22] and compounds such as 2,3-butanedione [23] and furan, 2-pentyl [24] have been linked to grain-based diets.

Therefore, the overall purpose of the present experiment was to evaluate the influence of the production system (intensive and extensive) on the chemical composition and the volatile profile in the muscle *longissimus thoracis et lumborum* of BEDM breed lambs.

2. Materials and Methods

2.1. Lamb Rearing and Feeding

In the present study, 28 lambs (males) of the Bordaleira-de-Entre-Douro-e-Minho (BEDM) sheep breed were raised for meat production in the Atlantic bioregion of Ponte de Lima (at Ponte Lima Agrarian School) under two different exploitation regimes: intensive and extensive system. Lambs were randomly selected from the flock and all of them were born and raised single. The weight of the lambs reared in the intensive production system at birth was 2.57 ± 0.28 kg, while those reared in the extensive regiment was 2.45 ± 0.27 kg, with no significant differences (p = 0.278) in the initial weights at the beginning of the experiment. In both farms, the feeding system was based on semi-natural pastures improved by sowing perennial ryegrass (*Lolium perenne*). The pastures were mainly constituted by grasses (54.3%) and legumes (28.9%). Specifically, 15 BEDM lambs were reared in the fall of 2018 under the intensive system and 13 BEDM lambs were reared in the spring of 2019 in the extensive system. Animals reared under the intensive system remained with the mothers and had ad libitum access to natural grass hay from birth to 3 months of age. After weaning (3 months), the lambs continued to be fed natural grass hay, in addition to 300 g/day of commercial compound feed supplied in two intakes per day (9:00 a.m. and 5:00 p.m.). The commercial compound feed used in the diet of intensively-reared lambs of the present research was supplied by Alimentação Animal Nanta S.A. (Marco de Canaveses, Portugal) and it was composed (in unknown proportions) of barley, wheat bran, extruded dehulled soy meal, dry beet pulp, brewers' dried grains, soy hulls, beet molasses, wheat germ, calcium carbonate, sunflower seed meal (extracted), soy oil, sodium chloride and a vitamins and minerals mix. Its chemical composition was the following: protein: 15.5%; ether extract: 4.5%; fiber: 8.2%; ash: 8.2%; calcium: 1.1%; phosphorous: 0.40%; sodium: 0.37%. All the information on the composition and ingredients of the commercial compound feed can be found in Supplementary Table S1. On the other hand, the lambs reared under the extensive system had access to their mother's milk (unweaned) and they went out to graze (ad libitum) with the herd from morning until dark during the entire experiment (from birth to slaughter; about 4 months). Upon darkness, the lambs were sheltered in stables where they also had access to meadow hay and water ad libitum. The growth test was carried out for 4 months and so the phenological status of the pasture was very varied. This test aimed to characterize the production systems in a holistic perspective; thus, the animals' feed was the one usually used in the farms.

2.2. Lamb Meat Samples

The trial planned to slaughter the animals at 4 months of age. Thus, the age at slaughter varied between 4 and 4.5 months and the births were not synchronized, which translated into the variation in the age at slaughter. With this in mind, at 4–4.5 months old, the lambs were transported to a commercial abattoir of Portugal. The animals were handled in batches ranging from 5 to 12 lambs and they were slaughtered according to the conditions previously reported [9]. Lambs reared in an intensive production system had a live weight of 13.54 ± 1.48 Kg (5.93 ± 1.02 Kg hot carcass weight), while those reared in an extensive production system had a live weight of 12.44 ± 2.65 kg (8.29 ± 1.77 Kg hot carcass weight). The live weight between both groups did not show significant differences ($p = 0.176$, while carcass weight of extensively-reared animals was significantly higher ($p < 0.01$) than those reared in the intensive production system. After cooling, the *longissimus thoracis et lumborum* muscles were removed from the sixth to the thirteenth vertebrae of lamb carcasses (a total of 56 pieces, 2 muscles, left and right and for each carcass). All muscle pieces were vacuum packed, refrigerated and transported to the CTC lab for the analysis. The left side was used for proximate composition analysis (72 h post-slaughter), while volatile analysis were carried out in the right muscle after refrigerated storage (4 ± 1 °C for 15 days). Before the analysis, a steak of each muscle (about 80 g) was conveniently chopped and homogenized in order to obtain a representative sample of each animal.

2.3. Analysis of Chemical Composition

Moisture [25], protein (Kjeldahl $N \times 6.25$) [26] and ash [27] were determined and expressed as percentage following the ISO recommended standards, while intramuscular fat was quantified according to the American Oil Chemistry Society (AOCS) official procedure [28].

2.4. Volatile Compounds Analysis

For the volatile compound analysis, Headspace-Solid phase microextraction (HS-SPME) technique was used for the volatile extraction and concentration, while the separation and identification of each volatile was carried out using gas chromatography coupled with mass spectrometry (GC-MS) (Agilent Technologies, Santa Clara, CA, USA) equipped with the DB-624 capillary column (30 m, 250 µm i.d., 1.4 µm film thickness; J&W Scientific, Folsom, CA, USA). All analysis steps, chromatographic and mass spectrometer conditions and data processing were previously reported [29]. The results were expressed as area units of the extracted ion chromatogram from the quantifier ion (m/z) per gram of sample (AU $\times 10^4$/g of sample). The Linear Retention Index (LRI) was calculated for the aforementioned capillary column (DB-624). Both LRI and m/z values are presented in all volatile tables as additional information to the volatile analysis.

2.5. Statistical Analysis

A total of 56 samples (28 for the chemical analysis and 28 for the volatile compounds determination) were analyzed in triplicate for each parameter. Normal distribution and variance homogeneity had been previously tested (Shapiro-Wilk). The influence of the production system on the chemical composition and volatile compounds was evaluated with one-way analysis of variance (one-way ANOVA) using the SPSS package version 23.0 (IBM SPSS, Chicago, IL, USA). Significant differences were indicated at $p < 0.05$, $p < 0.01$ and $p < 0.001$. Furthermore, the Pearson's linear coefficient was employed to determine correlations between the intramuscular fat and volatile content using the same statistical software.

3. Results and Discussion

3.1. Chemical Composition

The proximate composition of the BEDM lamb meat from the different production systems is shown as percentage in Table 1.

Table 1. Effects of the production system on the proximate composition of BEDM lamb *longissimus thoracis et lumborum* muscle.

	Intensive	Extensive	SEM	Sig.
Moisture (%)	78.00	75.91	0.290	***
Intramuscular fat (%)	0.49	1.51	0.132	***
Protein (%)	19.32	20.92	0.234	***
Ash (%)	1.37	1.20	0.022	***

SEM: Standard error of the mean. Sig.: Significance. *** ($p < 0.001$).

The values found for the proximate composition parameters agree with those reported by other authors. In this regard, a recent study comparing three different lamb breeds found values for fat (about 1.6%), protein (19–21%), moisture (75–77%) and ash (1.06–1.22%) and are similar to those described in this study [30]. Similarly, an investigation studying the influence of five different breeds and three (intensive, semi-extensive and extensive) production systems [9] or the influence of different slaughtered ages also showed comparable values for all proximate parameters.

As it can be observed, the production system significantly ($p < 0.001$) affected all the composition parameters. Concretely, the extensive production system provided lambs with a significantly ($p < 0.001$) higher intramuscular fat (IMF) and protein content than the intensive production system (1.51 vs. 0.49% and 20.92% vs. 19.32%, respectively). In contrast, intensively-reared lambs showed significantly ($p < 0.001$) higher amounts of moisture and ash (78.00 vs. 75.91% and 1.37 vs. 1.20%, respectively). Our results agree with those reported by other authors who observed that lamb meat with the highest moisture content presented the lowest IMF and protein contents [31]. Thus, inverse correlation between the moisture and IMF contents previously described in the lamb meat [9,31] explain our findings. However, these differences do not remain constant throughout the literature. Other studies found that grass-fed lambs decreased intramuscular fat [32,33] and protein content [34] while the moisture percentage was increased [34]. Several authors even observed that not all composition parameters were affected by the diet [35–37]. Among all proximate composition parameters, intramuscular fat is an important parameter that influenced the lamb meat quality. However, there is controversy about the influence of multiple factors on this content. In this regard, a recent study demonstrated that rearing season had an important effect on IMF content [38]. Sheep reared in spring presented higher IMF content than those reared in autumn. This fact could partially explain the results obtained by us, since the lambs reared in the extensive system (spring of 2019) presented higher values of IMF than those reared in the intensive system (fall of 2018). The differences in the availability and the quality of pasture could be an important factor that could explain the fact that animals reared in spring presented higher IMF than those reared during the autumn, since the two main peaks in lamb feeding change are in winter and spring [38]. Additionally, the better quality of the pasture also results in a better milk production by the mothers characterized by a high fat content due to a diet rich in fiber, which is undoubtedly related to the higher IMF content in the animals raised in the extensive system (unweaned) than those raised in the intensive production system (weaned). In line with the aforementioned elements, another important factor that influences IMF content is the diet. Generally speaking, the lambs feeding with concentrate presented higher IMF than those feeding with pasture or silage. This fact was corroborated by Cadavez et al. [9], who reported that the lambs reared in intensive production systems had higher IMF than those reared in semi-extensive or extensive systems. This is related with the fact that feedlot lambs had lower energy expenditure for grazing than lambs reared in the extensive system [39]. However, as reported in the Material and Methods section, in the present study both groups of animals graze and, thus, in our study we expect similar expenditure for grazing in animals from both production systems. Contrary to the results reported by Cadavez et al. [9], a study in which lambs received silage, silage + concentrate or concentrate during 36, 54 or 72 days concluded that both diet and feeding durations did not have an effect on IMF [34]. They attributed the lack of differences to the similarity in energy expenditure between animals and a higher rate of gain from good quality grass. The administration of the different amounts of concentrate in the diet, as well as the slaughter weight were parameters that did not affect the IMF in Barbarine lambs [40]. Similarly, in another study comparing lambs feeding with pasture and those that are stall-fed also found no significant differences on IMF between the groups [39]. Other authors reported that the weaned treatment (early, middle and unweaned) did not influence the IMF [41]. In contrast, in our case, the extensively-reared lambs (unweaned) presented higher IMF than the intensively-reared lambs (weaned at 3 months age). This fact could partially explain the differences of IMF between groups, since a previous meta-analysis study demonstrated that lambs that received milk had higher IMF than those that only had access to the pasture alone [42]. In Addition, the weaning also affected the carcass weight, since the unweaned lambs had heavier carcasses (both under concentrate and pasture feeding regimes) than the weaned animals [43]. This result agrees perfectly with our findings, since animals reared in the extensive systems (unweaned) presented both higher IMF and higher carcass weight than lambs reared in the intensive production system. Moreover, despite the fact that the carcass weight was significantly higher in extensively-reared animals, the live weight at slaughter did not show significant differences between both treatments (13.54 vs. 12.44 kg for intensively-reared and extensively-reared lambs, respectively). Similar results were observed in the research of Boughalmi and Araba [44], who found that the feeding management system (grazing vs. grazing with supplement vs. concentrate diet) did not affected the live weight of Timahdite lambs. In another research and in accordance with our results, the authors observed that the grass-fed lambs presented higher values of IMF (2.4% vs. 1.4%) than lambs offered the concentrate diet [45]. In this case, the authors attributed this fact to the adaptation period after weaning to the indoor condition and the change of diet type, which could also explain the results found by us in the present study.

Nevertheless, the large differences found in the literature may be due to the distinct conditions of the studies (age and weight of slaughter, the diet composition, management, breed, gender, etc.). Some authors reported, in the same study, contrary behavior of IMF content between two breeds feeding with three systems [46]. In this case, the authors reported that Akkaraman lambs feeding with concentrate presented lower values of IMF than those that received pasture, while in the Anatolian Merino lambs the concentrate-feeding lambs presented the highest IMF content [46]. This demonstrated that multiple factors could affect this parameter. In fact, in a recent study, the authors reported that IMF is strongly affected by diet, sex and age [47]. Thus, it is difficult to attribute the differences in IMF values to a single factor. However, in the present study, the IMF differences could be attributable to the different rearing season of the animal groups (availability and quality of pasture), the weaning treatment and also due to the adaptation period of lambs to concentrate diet.

3.2. Volatile Profile

In this research, a total of 205 volatile compounds from *longissimus thoracis et lumborum* of the BEDM breed were identified in the headspace of raw meat employing the SPME/GC-MS technique. The compounds obtained were divided into nine families according to their chemical nature: hydrocarbons (linear, branched, aromatic and benzene-derived hydrocarbons), acids, alcohols, aldehydes, ketones, esters, ethers, furans and sulfur compounds.

3.2.1. Hydrocarbons: Linear, Branched, Cyclic and Benzene-Derived

Table 2 displays the influence of the production system on the different hydrocarbons of the raw lamb meat. A total of 99 compounds belonging to this group were found, 70 in intensive-reared lambs and 48 in extensive-reared lambs. Concretely, in intensively-reared animals the hydrocarbons were distributed as follows: 9 linear hydrocarbons, 41 branched hydrocarbons, 16 cyclic hydrocarbons and 4 benzene-derived hydrocarbons. On the other hand, in extensively-reared lambs the hydrocarbons consisted of 11 linear hydrocarbons, 24 branched hydrocarbons, 11 cyclic hydrocarbons and 2 benzene-derived hydrocarbons.

Table 2. Effects of the production system on hydrocarbons (expressed as AU $\times 10^4$/g fresh weight) of BEDM lamb *longissimus thoracis et lumborum* muscle.

	LRI	m/z	Intensive	Extensive	SEM	Sig.
Linear hydrocarbons						
Butane	496	43	0.00	3.68	0.480	***
Pentane	500	43	9.52	11.75	0.961	ns
Heptane	700	71	1.00	1.29	0.100	ns
Octane	800	85	0.00	8.49	0.884	***
4-Octene, (E)-	841	55	0.00	2.42	0.241	***
Decane	1000	57	3.45	99.48	10.190	***
Undecane	1100	57	5.53	0.27	0.633	***
1-Undecene	1129	83	0.68	0.48	0.057	ns
Dodecane	1200	57	3.33	1.30	0.263	***
Hexadecane	1210	57	0.00	1.30	0.139	***
1-Tetradecene	1260	71	0.23	0.00	0.030	***
Tridecane	1300	57	1.55	0.47	0.146	***
Tetradecane	1400	57	0.87	0.00	0.101	***
Total linear hydrocarbons			26.15	130.93	11.279	***
Branched hydrocarbons						
Pentane, 2-methyl-	541	71	0.51	2.45	0.199	***
Pentane, 3-methyl-	550	56	1.18	29.79	2.839	***
Butane, 2,2,3,3-tetramethyl-	656	57	0.00	10.92	1.208	***
Hexane, 2,2-dimethyl-	656	57	17.00	0.00	2.111	***
Pentane, 2,3-dimethyl-	675	56	0.66	0.00	0.091	***
Pentane, 2,3,4-trimethyl-	759	71	21.00	0.09	2.375	***
Pentane, 2,3,3-trimethyl-	767	70	46.63	0.19	5.061	***
Pentane, 3-ethyl-	774	70	0.00	0.48	0.060	***
Hexane, 2,3-dimethyl-	774	70	1.45	0.00	0.214	***
1-Pentene, 3-ethyl-2-methyl-	778	55	1.37	0.00	0.169	***
3,4-Dimethyl-2-hexene	778	83	1.38	0.00	0.193	***
1-Pentene, 4,4-dimethyl-	788	57	0.00	0.61	0.084	***
Butane, 2,2,3-trimethyl-	789	85	0.56	0.00	0.067	***
Hexane, 2,2,5-trimethyl-	806	57	18.43	0.37	2.071	***
Heptane, 3-methylene-	820	70	5.10	0.00	0.692	***
Heptane, 3,4,5-trimethyl-	850	85	0.00	8.92	0.894	***
Pentane, 2,3,3,4-tetramethyl-	850	84	0.00	1.63	0.191	***
Heptane, 2,3-dimethyl-	850	85	1.42	0.00	0.180	***
Heptane, 2,6-dimethyl-	863	88	0.37	0.00	0.043	***

Table 2. *Cont.*

	LRI	*m/z*	Intensive	Extensive	SEM	Sig.
Heptane, 3-ethyl-	917	57	1.50	0.00	0.166	***
Nonane, 3,7-dimethyl-	925	57	1.08	0.00	0.129	***
Heptane, 2,2,4-trimethyl-	933	57	1.99	1.20	0.168	*
Heptane, 3,3,5-trimethyl-	947	71	0.00	0.46	0.048	***
Octane, 3,3-dimethyl-	947	71	1.95	0.00	0.218	***
Hexane, 2,3,4-trimethyl-	948	57	0.00	0.43	0.046	***
Pentane, 2,2-dimethyl-	948	57	1.29	0.00	0.141	***
3-Ethyl-2-methyl-1-heptene	996	84	0.89	0.00	0.100	***
Heptane, 3-ethyl-5-methylene-	998	70	0.00	1.94	0.226	***
2,3-Dimethyl-1-hexene	1037	55	1.80	0.00	0.191	***
Pentane, 3,3-dimethyl-	1046	71	3.01	0.00	0.349	***
1-Hexene, 3-methyl-	1062	70	3.04	0.00	0.386	***
(Z)-4-Methyl-2-hexene	1072	98	0.89	0.00	0.096	***
2,2,4,4-Tetramethyloctane	1078	57	146.91	17.31	16.485	***
1-Hexene, 5,5-dimethyl-	1090	57	0.00	58.12	6.290	***
Nonane, 5-butyl-	1097	127	1.45	0.00	0.172	***
Nonane, 5-(2-methylpropyl)-	1097	71	7.01	0.00	0.868	***
Heptane, 2,3,4-trimethyl-	1097	57	0.00	57.47	6.317	***
Dodecane, 2,6,10-trimethyl-	1097	57	14.27	0.00	1.805	***
Heptane, 2,2-dimethyl-	1101	57	0.93	0.00	0.141	***
Decane, 6-ethyl-2-methyl-	1104	57	0.00	83.25	8.323	***
Heptane, 3,3,4-trimethyl-	1135	71	0.00	0.50	0.060	***
Nonane, 2-methyl-	1136	57	0.69	0.00	0.084	***
Hexane, 1-(hexyloxy)-3-methyl-	1147	57	2.34	0.00	0.275	***
2-Undecene, 9-methyl-, (Z)-	1152	98	2.66	0.00	0.286	***
4-Undecene, 5-methyl-	1165	168	0.30	0.00	0.036	***
Pentane, 3,3-diethyl-	1181	98	0.34	0.00	0.036	***
2-Undecene, 3-methyl-, (Z)-	1203	70	0.60	0.00	0.065	***
Octane, 2,4,6-trimethyl-	1210	71	0.00	0.86	0.089	***
5-Ethyl-1-nonene	1224	83	0.32	0.00	0.039	***
1-Decene, 2,4-dimethyl-	1224	70	0.42	0.00	0.052	***
Hexane, 2-methyl-4-methylene-	1227	71	0.00	0.42	0.046	***
Heptadecane, 8-methyl-	1227	71	1.12	0.00	0.142	***
Undecane, 5-ethyl-	1242	57	1.09	0.00	0.157	***
Dodecane, 2-methyl-	1257	57	0.22	0.00	0.029	***
1-Undecene, 8-methyl-	1260	97	0.34	0.00	0.046	***
Tridecane, 3-methyl-	1331	57	0.00	0.39	0.039	***
Heptane, 2,4-dimethyl-	1349	71	0.00	0.35	0.036	***
5,5-Dibutylnonane	1358	71	0.00	0.36	0.037	***
Total branched hydrocarbons			315.54	278.51	15.461	ns

Table 2. *Cont.*

	LRI	*m/z*	Intensive	Extensive	SEM	Sig.
Cyclic hydrocarbons						
Cyclopentane, 1,2-dimethyl-, cis-	666	56	0.49	3.39	0.336	***
Cyclohexane, methyl-	720	83	0.00	5.77	0.628	***
Bicyclo[3.2.0]hepta-2,6-diene	810	91	15.94	12.09	0.788	*
Cyclopentane, 1,2,3-trimethyl-	820	56	0.62	0.00	0.083	***
Cyclooctane	822	70	0.00	2.75	0.275	***
Cyclohexane, 1,3-dimethyl-, cis-	840	97	2.02	0.00	0.228	***
Cyclohexane, 1,3-dimethyl-	840	97	0.49	0.00	0.059	***
Cyclobutane, 1,1,2,3,3-pentamethyl-	938	70	1.69	0.00	0.173	***
Cyclopropane, 1-methyl-2-pentyl-	942	55	0.34	0.00	0.037	***
Bicyclo[3.1.1]hept-2-ene, 3,6,6-trimethyl-	992	93	2.74	0.00	0.289	***
Cyclopentane, 1,2,3,4,5-pentamethyl-	996	69	0.86	0.00	0.094	***
Cyclohexane, butylidene-	1042	67	0.00	0.72	0.078	***
Cyclodecene, (Z)-	1042	67	3.64	0.00	0.371	***
Cyclopropane	1063	41	3.04	0.00	0.318	***
Cyclohexane, 1,2-diethyl-1-methyl-	1075	125	0.52	0.00	0.057	***
Cyclopentane, pentyl-	1084	68	1.87	0.00	0.193	***
D-Limonene	1085	93	0.00	0.90	0.099	***
Cyclooctane, methyl-	1129	55	0.00	0.74	0.115	***
Cyclopentane, 1-ethyl-1-methyl-	1143	83	0.00	2.05	0.216	***
Butane, 2-cyclopropyl-	1165	70	0.98	0.00	0.116	***
Cyclododecane	1249	83	0.62	0.00	0.072	***
Heptylcyclohexane	1322	82	0.95	0.78	0.094	ns
Cyclopropane, 1,1,2,3-tetramethyl-	1374	71	0.00	0.46	0.051	***
Cyclohexane, octyl-	1444	82	0.00	0.24	0.024	***
Total cyclic hydrocarbons			36.81	29.89	0.980	***
Benzene-derived hydrocarbons						
Ethylbenzene	928	91	0.82	0.00	0.087	***
Benzene, 1,3-dimethyl-	937	106	2.54	1.58	0.177	**
Benzene, n-butyl-	1118	91	0.91	0.00	0.097	***
Benzene, (1,1-dimethylethoxy)-	1137	94	3.24	0.54	0.286	***
Total benzene-derived hydrocarbons			7.50	2.12	0.590	***
TOTAL HYDROCARBONS			386.00	441.44	18.278	ns

SEM: Standard error of the mean. Sig.: Significance. * ($p < 0.05$); ** ($p < 0.01$); *** ($p < 0.001$); ns: no significant difference.

As can be observed, the production system did not significantly affect the total hydrocarbon content, although this was slightly higher in lambs produced under extensive conditions (441.44 vs. 386.00 AU $\times$ 10^4/g fresh meat). However, the total value of families of linear, cyclic and benzene-derived hydrocarbons were significantly ($p < 0.001$) affected by the production system. Specifically, lambs reared in the extensive system presented higher amounts of total linear hydrocarbons (130.93 vs. 26.15 AU $\times$ 10^4/g fresh meat). On the contrary, lambs reared intensively had significantly ($p < 0.001$) higher concentrations of total cyclic hydrocarbons (36.81 and 29.89 AU $\times$ 10^4/g fresh meat for intensive and extensive systems, respectively) and benzene-derived hydrocarbons (7.50 AU $\times 10^4$/g fresh meat for intensive and 2.12 AU $\times 10^4$/g fresh meat for extensive and 2.12). Numerically but not significantly, the branched hydrocarbons from intensively-reared animals were also higher (315.54 and 278.51 AU $\times 10^4$/g fresh meat for intensive and extensive production systems, respectively).

Most individual hydrocarbons were significantly ($p < 0.05$) affected by the production system apart from pentane, heptane, 1-undecene and heptylcyclohexane. Differences in volatile compounds attributed to the production system may arise from the origin of the animal feed, since some of these volatile compounds, such alkanes of more than 10 carbons, can be stored in fatty tissues through diet [48,49]. However, the individual trends varied depending on the substance in question. Thus, in intensively-reared lambs, the linear hydrocarbon that was found in the highest concentration was pentane (9.52 AU $\times 10^4$/g fresh meat), while for extensively-reared lambs it was decane (99.48 AU $\times 10^4$/g fresh meat). In the case of branched hydrocarbons, the highlights were 2,2,4,4-tetramethyloctane, with a concentration of 146.91 AU $\times 10^4$/g fresh meat, and heptadecane, with a concentration of 83.25 AU $\times 10^4$/g fresh meat, for intensive-raised and extensive-raised lambs, respectively. Moreover, for both production systems, the cyclic hydrocarbon with the highest presence was the same (namely bicyclo[3.2.0]hepta-2,6-diene) and shoed concentrations of 15.94 and 12.09 10 AU $\times 10^4$/g fresh meat for lambs reared in intensive and extensive systems, respectively. Within this group of hydrocarbons, it is also worth highlighting the presence of the terpene D-limonene in grass-fed lambs (0.90 AU $\times 10^4$/g fresh meat) and its absence in lambs fed with concentrate.

Additionally, it should be noted that benzene-derived hydrocarbons did show the same trend since all the compounds belonging to this group (namely ethylbenzene; benzene, 1,3-dimethyl-; benzene, n-butyl-; benzene, (1,1-dimethylethoxy)-) were found in significantly ($p < 0.01$) higher concentrations in lambs fed under the intensive production system. These results are in disagreement with those obtained by various authors who reported that benzene-derived hydrocarbons were produced to a greater extent in lambs fed by grazing than by concentrate [6,17,50]. This discrepancy is difficult to explain since normally benzene-derived hydrocarbons are related to the consumption of grass and, more specifically, to the carotenoids present in green plants [51] or even with the contaminants retained by these vegetables [52,53].

On the other hand, hydrocarbons constituted the largest family of volatile compounds detected in intensive and extensive systems (59.78% and 54.76%, respectively), with branched hydrocarbons being the volatile compounds most abundant in both diets (48.87% for lambs reared in intensive production system and 34.55% for animals reared in extensive production system) and benzene-derivatives being the least abundant ones (1.16 and 0.26% for intensively-reared and extensively-reared lambs, respectively) (Figure 1). In spite of the distributions of these percentages, linear hydrocarbons (which represent 4.05 and 16.24% in intensively-reared and extensively-reared lamb meat, respectively) and cyclic hydrocarbons (5.70% for intensively-reared and 3.71% for extensively-reared lambs) taken together with branched ones, in general, are not particularly important in contributing to the aroma of meat as they have high odor thresholds [29,54–56]. On the contrary, benzene-derived hydrocarbons, even those possessing a low percentage of the total volatile content, could have a significant contribution to the volatile pattern of lamb meat due to their low odor threshold [48,56,57].

3.2.2. Acids

Seven acids were identified in the meat of BEDM lambs, four in samples from the intensive production system and six from the extensive system. Moreover, the production system significantly ($p < 0.05$) affected both the total amount of acids and that of each individual compound (Table 3). Specifically, extensively-reared lambs showed a higher concentration of all the acids determined with the exception of hexanoic acid, which appeared in intensively-reared lambs (1.33 AU $\times 10^4$/g fresh meat) while it was not detected in extensive farming lambs. In addition, the total amount of acids was also significantly ($p < 0.001$) higher in the lambs reared extensively (8.87 vs. 2.51 AU × 10^4/g fresh meat).

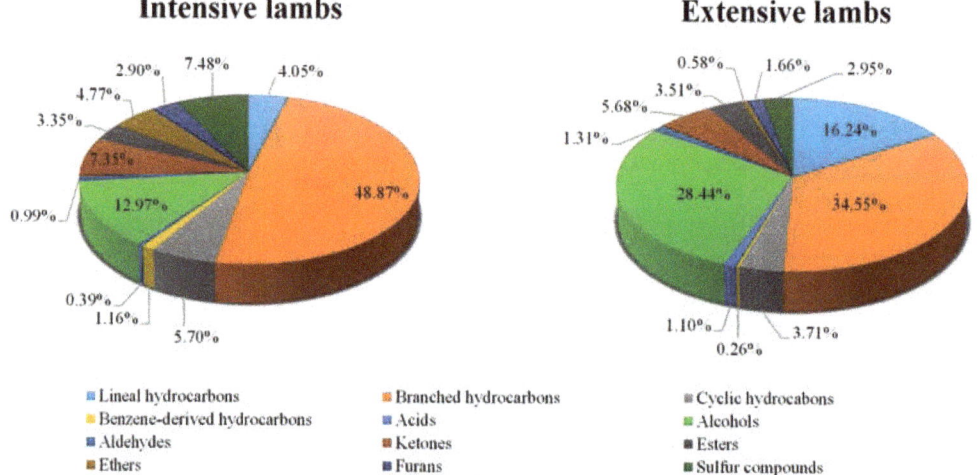

Figure 1. Volatile families of BEDM lamb *longissimus thoracis et lumborum* muscle (expressed as percentages) affected by the production system.

Table 3. Effects of the production system on acids and alcohols (expressed as AU × 10^4/g fresh weight) of BEDM lamb *longissimus thoracis et lumborum* muscle.

	LRI	m/z	Intensive	Extensive	SEM	Sig.
Acids						
Acetic acid	696	60	0.05	0.45	0.044	***
2-Propenoic acid	709	55	0.00	3.64	0.393	***
Butanoic acid	929	60	1.10	1.91	0.164	*
Pentanoic acid	1101	60	0.00	1.67	0.204	***
Hexanoic acid	1102	60	1.33	0.00	0.158	***
Pentanoic acid, 2-methyl-, anhydride	1157	99	0.04	0.95	0.106	***
Nonanoic acid	1314	60	0.00	0.26	0.031	***
Total acids			2.51	8.87	0.678	***
Alcohols						
Glycidol	499	44	2.10	90.02	12.230	***
1-Propanol	570	59	0.20	1.05	0.100	***
1-Butanol	709	56	2.01	29.94	3.053	***
1-Butanol, 3-methyl-	814	55	0.22	1.89	0.199	***
1-Butanol, 2-methyl-	818	57	0.00	4.70	0.506	***
1-Pentanol	855	55	0.00	33.61	3.542	***
Cyclobutanol, 2-ethyl-	875	56	1.03	0.00	0.130	***
2-Octen-1-ol, (Z)-	875	67	0.72	0.00	0.098	***
2,3-Butanediol, [S-(R*,R*)]-	929	45	3.44	0.00	0.408	***

Table 3. Cont.

	LRI	m/z	Intensive	Extensive	SEM	Sig.
DL-2,3-Butanediol	931	45	0.00	0.73	0.084	***
1-Butanol, 3-methyl-, acetate	952	55	0.05	1.37	0.306	*
1-Hexanol	967	55	3.70	7.84	0.632	***
1-Heptanol	1062	70	4.66	5.29	0.449	ns
1-Octen-3-ol	1068	57	39.65	33.11	3.525	ns
Ethanol, pentamethyl-	1079	59	0.00	0.72	0.074	***
2,3,4-Trimethyl-1-pentanol	1099	71	6.45	0.00	0.795	***
1-Hexanol, 2-ethyl-	1113	57	4.91	2.62	0.371	***
1-Hexanol, 5-methyl-2-(1-methylethyl)-	1128	71	0.94	0.00	0.109	***
1-Undecanol	1129	69	0.00	0.32	0.035	***
4-Ethylcyclohexanol	1130	81	0.24	0.41	0.052	ns
Benzyl alcohol	1145	108	0.27	0.00	0.030	***
5-Methyl-1-heptanol	1143	70	1.11	2.70	0.220	***
1-Octanol	1147	56	3.25	3.95	0.289	ns
2-Octen-1-ol, (E)-	1148	57	1.62	1.94	0.203	ns
3-Octen-2-ol, (E)-	1148	67	0.00	1.02	0.117	***
3-Octen-1-ol, (Z)-	1149	81	0.69	0.00	0.116	**
1-Butanol, 2-methyl-, trifluoroacetate	1152	70	3.35	0.00	0.361	***
1,8-Octanediol	1168	55	0.00	4.04	0.562	***
6-Undecanol	1183	55	0.00	0.78	0.093	***
4-Methyl-5-decanol	1184	83	0.42	0.00	0.066	***
1-Butanol, 3,3-dimethyl-	1189	56	0.00	0.37	0.038	***
1,9-Nonanediol	1224	55	0.00	0.20	0.021	***
1-Nonanol	1224	56	0.21	0.15	0.014	*
1-Butanol, 2-methyl-, propanoate	1349	57	0.00	0.52	0.053	***
2,4-Di-tert-butylphenol	1456	191	2.52	0.00	0.353	***
Total alcohols			83.76	229.30	16.597	***

SEM: Standard error of the mean. Sig.: Significance. * ($p < 0.05$); ** ($p < 0.01$); *** ($p < 0.001$); ns: no significant difference.

Within this group, branched chain fatty acids, such as 4-methyloctanoic, 4-ethyloctanoic and 4-methylnonanoic acids, are of special interest because they are related to the specific aroma of lamb meat, contributing to the mutton-like aroma [6,58,59]. However, none of these compounds were detected in the lambs analyzed regardless of the production system employed.

The contribution of acids on the total volatile compounds was very low. Indeed, this family has been the least abundant in intensively-reared lambs and the second with the least presence in extensively-reared lambs. More concretely, total acids represented 0.39 and 1.10% of the total volatile substances in lambs reared in intensive and extensive production systems, respectively (Figure 1). This weak presence may be due to the fact that some acids, such as branched chain, are found mainly in adipose tissue since they are diminished in muscle tissue [58]. Furthermore, since branched fatty acids tend to increase with the age of the animals and are associated with older lambs of over two years [60], their presence in our study was limited due to the young age of the lambs (~4 months).

3.2.3. Alcohols

In this study, 35 different alcohols (24 in animals from the intensive production system and 25 in animals from the extensive production system) were detected in the BEDM lambs meat (Table 3). As can be observed, all these compounds were significantly ($p < 0.05$) affected by the production system except for 1-heptanol, 1-octen-3-ol, 4-ethylcyclohexanol, 1-octanol and 2-octen-1-ol, (E)-, although in different manners. Nevertheless, it can be generally observed that the extensive production system tends to provide BEDM lamb meat with a higher presence of alcohols, since 20 of the compounds identified in this group were found in significantly ($p < 0.05$) higher concentrations compared to their intensively-reared counterparts. Moreover, BEDM lambs reared in the extensive production system showed a total content of alcohols significantly ($p < 0.001$) higher than those reared in the intensive system (229.30 and 83.76 AU × 10^4/g fresh meat, respectively). This occurrence could be due to the fact that some alcohols, such as 1-pentanol and 1-hexanol, are related to the degradation of their homologous aldehydes during lipid oxidation [61,62]. In this regard, a previous study demonstrated that the BEDM lambs reared extensively have very high contents of polyunsaturated fatty acids (specially n-3 PUFA) [9], which are more susceptible to oxidation [63] and can explain the results observed on the lipid-derived volatile compounds behavior. Thus, in our study, 1-pentanol alone has been identified in extensively-reared lambs (33.61 AU × 10^4/g fresh meat) and 1-heptanol has shown a concentration of 7.84 AU × 10^4/g fresh meat in extensively-reared lambs compared to 3.70 AU × 10^4/g in intensively-reared lambs. Nevertheless, the greater levels of 1-hexanol in extensively-reared lambs contrasts with the fact that this alcohol comes from the autoxidation of linoleic acid [63,64], which is typically present in concentrates made from grains [65]. Despite these observations, other studies have also found that grass-raised ewes showed higher amounts of 1-hexanol in meat than intensively-reared ewes [18]. In addition, 1-pentanol and 1-hexanol could positively affect the aroma of lamb meat since 1-pentanol is characterized by its pleasant, sweet or fruity odor, while 1-hexanol has a herbal and fatty odor [29,48,66]. Furthermore, the lambs reared in the extensive production system displayed a concentration of 1-butanol that is significantly higher than that of the lambs reared intensively (29.94 vs. 2.01 AU × 10^4/g fresh meat). These results agree with the fact that meat from BDEM lambs reared in extensive system presented high amounts of linoleic acid [9], which is the main precursor of this volatile compound (derived from oxidation reactions) [63].

On the other hand, two alcohols (namely benzyl alcohol and 2,4-di-tert-butylphenol) have been identified, which could be related to the diet based on grass as they are phenolic compounds [19]. However, in the meat of BEDM lambs reared in extensive production system (fed with grass) none of these two compounds were identified, while in those fed with concentrate (intensive production system) values of 0.27 and 2.52 AU × 10^4/g fresh meat were obtained for benzyl alcohol and 2,4-di-tert-butylphenol, respectively. These results are similar to those indicated by other authors, which suggest that not all phenolic compounds are related to grass [14,67].

Regarding the 1-octanol, this alcohol was not significantly affected ($p > 0.05$) by breeding, obtaining very similar values for both types of lambs (3.25 and 3.95 AU × 10^4/g fresh meat for intensively-reared and extensively-reared lambs, respectively). Similarly, 1-octen-3-ol has not been significantly ($p > 0.05$) affected by the production system, since both lambs showed concentrations in the same range (39.65 and 33.1 AU × 10^4/g for intensively- and extensively-reared lambs, respectively). This could be due to the fact that 1-octen-3-ol is a compound that arises from several pathways [48]; it is a volatile substance derived from lipid oxidation that is frequently reported in meat and meat products [63,68]. These facts are in agreement with those obtained by Sivadier et al. [17] who observed that 1-octen-3-ol content did not depend on the diet supplied. In addition, although they are normally of lower molecular weight, there are various alcohols that are considered to be of metabolic origin; thus, they are not affected by the diet provided [69].

With respect to the contribution of alcohols over the total volatile content, this family is the second most abundant in both production systems. Specifically, this group represents 12.97 and 28.44% in the lambs reared intensively and extensively, respectively (Figure 1). Despite this, alcohols have a debatable high odor threshold and their contribution to volatile flavor is less than that of other compounds such as aldehydes [70]. However, various alcohols, such as 1-pentanol, may contribute to the lamb aroma on account of their low odor threshold and their mild, fruit and balsamic aroma [66,71].

3.2.4. Aldehydes

In our work, the lambs reared extensively showed a significantly ($p < 0.001$) higher total aldehyde value than lambs reared intensively (10.53 vs. 6.40 AU $\times 10^4$/g fresh meat), which is inconsistent with what was obtained in Almela et al. [5]. Within the aldehyde family, 14 compounds were identified (11 in lambs from intensive production and 8 in lambs from extensive production system) (Table 4), among which hexanal was the only volatile compound that was not significantly ($p > 0.05$) affected by the production system and similar concentrations in both groups of lambs were found (1.07 and 0.97 AU $\times 10^4$/g fresh meat). This suggests that the meat from both productions systems could show similar lipid oxidation states, since hexanal is assumed to be one of the main indicators of lipid oxidation [72,73]. However, other aliphatic saturated aldehydes found in our research, apart from hexanal, are also considered as indicators of lipid oxidation in raw meat because they are derived from the degradation of hydroperoxides [63,74]. This is the case of the octanal and nonanal aldehydes, which are derived from the oxidation of oleic acid [63,75]. Concretely, octanal was the one that was found in greater presence in the lambs reared under intensive conditions (2.30 AU $\times 10^4$/g fresh meat), while it did not appear in the lambs reared extensively. This fact could indicate that intensively-reared lambs may have a greater intensity of rancid odor, since previous studies have found that octanal is the aldehyde that presented the highest correlation with this parameter in lamb meat packed under a protective atmosphere [76]. On the contrary, nonanal was only detected in extensively-reared lambs since it is the aldehyde that appears in the highest concentration in the meat of these animals (2.56 AU $\times 10^4$/g fresh meat), after 2-propenal with a concentration of 4.04 10^4/g fresh meat, which would provide a plastic and soapy aroma [77].

Table 4. Effects of the production system on aldehydes and ketones (expressed as AU $\times 10^4$/g fresh weight) of BEDM lamb *longissimus thoracis et lumborum* muscle.

	LRI	m/z	Intensive	Extensive	SEM	Sig.
Aldehydes						
Propanal, 2-methyl-	556	72	0.00	0.21	0.021	***
Butanal, 3-methyl-	659	58	0.26	0.57	0.050	***
Butanal, 2-methyl-	671	57	0.14	1.00	0.097	***
2-Butenal	841	70	0.39	0.00	0.047	***
Hexanal	874	56	1.07	0.97	0.109	ns
Heptanal	987	70	0.27	0.95	0.100	***
Hexanal, 3-methyl-	988	55	0.23	0.00	0.034	***
Hexanal, 3,3-dimethyl-	1006	69	0.92	0.00	0.099	***
Octanal	1084	84	2.30	0.00	0.276	***
Benzeneacetaldehyde	1139	91	0.09	0.23	0.026	**
2-Propenal	1148	55	0.00	4.04	0.437	***
Nonanal	1168	57	0.00	2.56	0.278	***

Table 4. *Cont.*

	LRI	m/z	Intensive	Extensive	SEM	Sig.
2-Decenal, (E)-	1298	83	0.40	0.00	0.039	***
2-Decenal, (Z)-	1299	70	0.32	0.00	0.040	***
Total aldehydes			6.40	10.53	0.555	***
Ketones						
2,3-Butanedione	589	86	15.54	0.00	1.711	***
2-Butanone	593	72	0.72	1.93	0.138	***
2-Pentanone	724	86	0.23	0.66	0.047	***
3-Pentanone	735	57	6.65	0.00	0.909	***
2,3-Pentanedione	739	100	0.00	0.79	0.097	***
1,5-Heptadien-4-one, 3,3,6-trimethyl-	779	83	0.00	0.99	0.115	***
Cyclobutanone, 2,2,3-trimethyl-	815	70	0.00	2.90	0.306	***
3-Heptanone	973	57	0.32	2.32	0.258	***
2-Heptanone	980	58	1.68	8.67	0.851	***
Pyrolo[3,2-d]pyrimidin-2,4(1H,3H)-dione	1057	151	9.83	8.21	0.456	ns
3-Ethylcyclopentanone	1058	83	0.00	0.37	0.042	***
4-Octanone, 5-hydroxy-2,7-dimethyl-	1059	69	0.00	2.49	0.283	***
Butyrolactone	1061	86	2.71	0.00	0.283	***
4-Hexen-3-one, 5-methyl-	1062	83	0.41	0.00	0.050	***
3-Heptanone, 5-methyl-	1069	99	3.28	0.00	0.395	***
5-Hepten-2-one, 6-methyl-	1073	68	0.74	0.49	0.059	*
2-Octanone	1077	58	2.16	4.08	0.334	*
2(3H)-Furanone, dihydro-5-methyl-	1095	56	0.00	8.53	1.108	***
5-Hexen-3-one	1151	98	0.90	1.24	0.076	*
3-Nonanone	1155	72	0.61	0.57	0.050	ns
2-Nonanone	1161	58	0.86	0.65	0.039	**
2(3H)-Furanone, 5-ethyldihydro-	1179	85	0.48	0.56	0.040	ns
2-Undecanone	1310	58	0.36	0.00	0.038	***
2(3H)-Furanone, dihydro-5-pentyl-	1400	85	0.00	0.36	0.037	***
Total ketones			47.48	45.81	10.626	ns

SEM: Standard error of the mean. Sig.: Significance. * ($p < 0.05$); ** ($p < 0.01$); *** ($p < 0.001$); ns: no significant difference.

Another important aldehyde is heptanal, which is usually an indicator of animal diets rich in linoleic acid, since it is an aldehyde that appears after the oxidation of this fatty acid [64]. In this manner, it would be expected that the lambs reared in the intensive system would obtain higher concentrations of heptanal than those reared extensively because linoleic acid is typically present at high quantities in cereal grains [65]. Conversely, in our research, lambs from extensive production system displayed significantly ($p < 0.001$) higher amounts of heptanal than intensive-reared lambs (0.95 vs. 0.27 AU $\times 10^4$/g fresh meat). It is important to highlight that in a previous study, the BDEM lambs reared in extensive system also presented high amounts of this fatty acid, which explains our findings [9]. These results are consistent with those shown by Vasta et al. [18], who found that milk from grass-fed ewes had higher concentrations of heptanal than those fed a grain-based diet. Therefore, it is not easy to unambiguously link an aldehyde compound with a lamb feeding or production system [14].

On the other hand, the fraction corresponding to the group of aldehydes with respect to the total volatiles was very low in both groups (0.99 and 1.31%, for intensive-reared and extensive-reared lambs, respectively) (Figure 1). Specifically, it is the second and third group of volatile compounds that are the less abundant of the nine divisions in lambs produced in intensive and extensive systems, respectively. This fact is in disagreement with the results reported by other authors who found that the aldehyde family generally represents the main contributors to the volatile fraction extracted from ruminant meat [18,78]. Despite this discrepancy, aldehydes remain one of the most important volatile compounds because they are the main indicators of rancidity in meat due to their low odor threshold [79,80].

3.2.5. Ketones

A total of 24 ketones were identified in the BEDM lamb meat (17 in intensively-reared and 18 in extensively-reared lambs). As shown in Table 4, the production system did not significantly ($p > 0.05$) affect the total amount of ketones, although it was slightly higher in extensively-reared lambs (47.48 vs. 45.81 AU $\times 10^4$/g fresh meat). Despite the fact that the total content of this family was not affected by production system, each individual ketone showed significant ($p < 0.05$) differences according to the production system employed, with the exception of pyrolo[3,2-d]pyrimidin-2,4(1H,3H)-dione; 3-nonanone and 2(3H)-furanone, 5-ethyldihydro-, also known as γ-hexalactone, which could be related to the metabolism of the ruminants since certain ketones are considered to be of metabolic origin [69]. Conversely, there are ketones that are derived from the diet [18]. This is the case of 2,3-octanedione, which has been considered by several studies as a typical compound present in grass-fed animals meat [18,20]. However, the results obtained by Resconi et al. [81] and Gravador et al. [68] did not identify 2,3-octanedione in lambs regardless of their diet.

On the other hand, 2,3-butanedione (diacetyl), was linked with grain diets [23]. This event is in agreement with the results obtained in our work, since it has been observed that only lambs reared in the intensive system had 2,3-butanedione (15.54 AU $\times 10^4$/g fresh meat), while this diketone was not identified in extensively-reared animals. In fact, 2,3-butanedione also stands out for being the ketone that appears in greater abundance in lambs fed with concentrate. Additionally, the presence of 2-heptanone and 2-butanone are associated with grain-based diets [16,65]. Despite this, in our study, it was found that lambs reared extensively presented significantly ($p < 0.001$) higher amounts of 2-heptanone, 2-octanone and 2-butanone (8.67 vs. 1.68 AU $\times 10^4$/g fresh meat for 2-heptanone; 4.08 vs. 2.16 AU $\times 10^4$/g fresh meat for 2-octanone; and 1.93 vs. 0.72 AU $\times 10^4$/g fresh meat for 2-butanone) and even 2-heptanone, which is the ketone that was detected in greater abundance in these lambs. These unexpected outcomes are consistent with those obtained by Vasta et al. [18] who did not observe significant differences in this 2-ketones, yet did find slightly higher amounts in lambs fed with grass than with concentrate. Additionally, the high proportion of these ketones in animals reared in the extensive production system could be due to 2-ketones being derived from lipid oxidation [63] and BEDM lambs that are extensively-reared had the highest amounts of PUFA [9], which promotes their formation. Contrary, 2-nonanone was identified in a significantly higher concentration ($p < 0.01$) in intensively-reared lambs (0.86 and 0.65 AU $\times 10^4$/g fresh meat for intensively-reared and extensively-reared animals, respectively). Although the value of 2-nonane is significant higher in lambs from intensive systems than in lambs from extensive production system, it is important to mention that the difference of content between both groups of animals was less than those described for the aforementioned 2-ketones. This ketone (2-nonanone) possesses a "fatty, oily, fruity" odor and has previously been associated with a lamb flavor [82,83], which could indicate that lambs reared in intensive production system could show a stronger flavor linked to this compound.

Furthermore, it should be noted that up to four different lactones were identified in the lamb meat, namely butyrolactone; 2(3H)-furanone, dihydro-5-methyl-; 2(3H)-furanone, 5-ethyldihydro-; and 2(3H)-furanone, dihydro-5-pentyl. It has been previously pointed out that this type of lactones have been linked to grain-based diets [20,21] due to its higher content of oleic and linoleic acids compared to pasture [84]. This is because lactones arise from the corresponding hydroxy-fatty acids [85], which in turn are formed in the rumen by the oxidation of dietary oleic and linoleic acids [86]. However, in our work, only butyrolactone seemed to follow the trend expected, since it was found in intensively-reared lambs (2.71 AU $\times 10^4$/g fresh meat) and not in extensively-reared lambs. On the contrary, 2(3H)-furanone, dihydro-5-methyl- and 2(3H)-furanone, dihydro-5-pentyl- were only identified in lambs produced in extensive systems and obtained concentrations of 8.53 and 0.36 AU $\times 10^4$/g fresh meat in these lambs, respectively. Finally, 2(3H)-furanone, 5-ethyldihydro- was not significantly ($p > 0.05$) affected by the production system.

Regarding the contribution of ketones on the total volatile compounds, this family represented 7.35 and 5.68% of the total volatile substances in intensively-reared and extensively-reared lambs, respectively (Figure 1). This percentage was slightly lower than that reported by Krvavica et al. [87], who observed ketone values of around 9% in lamb of the Lika breed. Despite this, the percentage of ketones is relatively high, since this group is the fourth most abundant family for intensively-reared lambs and the third for extensively-reared lambs within the nine groups. This occurrence combined with the fact that ketones have a low perception threshold [56,82] renders this group a notable contributor to the meat flavor [73].

3.2.6. Esters, Ethers, Furans and Sulfur Compounds

Sixteen different esters were detected in BEDM lambs meat (8 in intensively-reared and 11 in extensively-reared lambs), which were significantly ($p < 0.001$) affected by the production system except for a single compound, namely 2-butenoic acid, 2-methyl-, 2-methylpropyl ester, which did not suffer significant ($p > 0.05$) variations (Table 5). In general, esters were found to a greater extent in lambs reared in extensive systems since 10 of the 16 compounds obtained significantly ($p < 0.001$) higher concentrations in these animals. In addition, the total content of esters was also significantly ($p < 0.05$) higher in the lambs reared in extensive systems compared to those reared in intensive systems (28.30 vs. 21.62 AU $\times 10^4$/g fresh meat). These differences could be related to the possible variability of the fatty acid profile of lambs [88] because the main origin of esters is the esterification of carboxylic acids [89]. Despite the differences, previous studies have shown that the contribution of esters to the aromatic profile of lamb meat may be low [68]; several authors did not even detect these compounds [17,81,90,91] or detected a low number of esters [37,76,87,88]. Therefore, although the fraction of esters to the total volatile compounds was relatively high (3.35 in intensively-reared lambs and 3.51% for lambs reared under extensive conditions) (Figure 1), their presence may not contribute to the overall aroma of the lamb meat.

Regarding the ethers group, only three different compounds were identified (Table 5). Two were found in intensively-reared lambs (namely, ether, 2-ethylhexyl tert-butyl and decyl heptyl ether) and one in lambs raised extensively (namely, ether, 3-butenyl pentyl). All these individual compounds as well as their total content were significantly ($p < 0.001$) affected by the production system. Specifically, the lambs fed under the intensive diet showed significantly ($p < 0.001$) higher amounts of this group (30.83 vs. 4.69 AU $\times 10^4$/g fresh meat). In addition, ethers represented 4.77% of the total volatile content in lambs from the intensive system and occupies the fifth position of the nine families, while this group only accounted for 0.58% of the total volatile content in lambs from the extensive system and is the family that appears in the lowest presence (Figure 1). The literature consulted did not frequently find these compounds in lamb meat and, in some cases, were non-existent in many investigations [18,34,91,92]. Furthermore, it was observed that ethers were not relevant compounds in the aroma of lambs [93] and some of these substances could be found in lamb due to their possible use as insecticides, acaricides and fumigants for the soil [48].

Table 5. Effects of the production system on esters, ethers, furans and sulfur compounds (expressed as AU $\times 10^4$/g fresh weight) of BEDM lamb *longissimus thoracis et lumborum* muscle.

	LRI	m/z	Intensive	Extensive	SEM	Sig.
Esters						
Acetic acid, methyl ester	537	74	0.18	0.46	0.044	***
Ethyl Acetate	598	43	0.64	4.17	0.439	***
Formic acid, ethenyl ester	708	43	0.00	11.24	1.193	***
Butanoic acid, ethyl ester	856	70	1.51	0.00	0.158	***
Formic acid, heptyl ester	1062	56	0.00	5.69	0.599	***
Sulfurous acid, 2-ethylhexyl nonyl ester	1086	57	15.82	0.00	1.848	***
Formic acid, octyl ester	1147	55	0.00	3.78	0.431	***
Propanoic acid, 2-methyl-, 2-propenyl ester	1177	71	0.00	0.55	0.070	***
Butanoic acid, 2-propenyl ester	1183	71	0.00	0.63	0.080	***
2-Butenoic acid, 2-methyl-, 2-methylpropyl ester	1183	83	0.31	0.45	0.039	ns
2-Propenoic acid, 2-methyl-, (tetrahydro-2-furanyl)methyl ester	1297	71	1.00	0.00	0.103	***
Sulfurous acid, hexyl nonyl ester	1298	85	1.70	0.00	0.182	***
Sulfurous acid, 2-ethylhexyl hexyl ester	1331	85	0.00	0.45	0.049	***
Propanoic acid, 2-methyl-, 2-methylpropyl ester	1384	71	0.00	0.38	0.046	***
Sulfurous acid, 2-ethylhexyl isohexyl ester	1412	57	0.46	0.00	0.054	***
Pentanoic acid, 5-hydroxy-, 2,4-di-t-butylphenyl esters	1454	191	0.00	0.49	0.060	***
Total esters			21.62	28.30	1.509	*
Ethers						
Ether, 3-butenyl pentyl	1046	55	0.00	4.69	0.532	***
Ether, 2-ethylhexyl tert-butyl	1090	57	28.67	0.00	3.068	***
Decyl heptyl ether	1169	57	2.15	0.00	0.271	***
Total ethers			30.83	4.69	2.849	***
Furans						
Furan, 2-ethyl-	706	81	0.90	4.75	0.468	***
Furan, 2,3-dihydro-	806	70	0.00	1.76	0.228	***
2-n-Butyl furan	956	81	0.33	0.47	0.042	ns
Furan, 2-pentyl-	1054	81	17.49	6.41	1.331	***
Total furans			18.73	13.38	1.059	**
Sulfur compounds						
Dimethyl sulfide	528	62	0.43	1.67	0.211	**
Carbon disulfide	532	76	47.57	21.34	4.129	***
Dimethyl sulfone	1090	79	0.30	0.81	0.091	**
Total sulfur compounds			48.30	23.81	4.010	**

SEM: Standard error of the mean. Sig.: Significance. * ($p < 0.05$); ** ($p < 0.01$); *** ($p < 0.001$); ns: no significant difference.

On the other hand, four furans were identified in both lambs (Table 5), except for furan, 2,3-dihydro-, which was only found in lambs reared in extensive systems at a concentration of 1.76 AU × 10^4/g fresh meat. Specifically, the furan that appeared in the highest concentration was furan, 2-pentyl in both production systems, which has been frequently identified in lamb meat [76,83,87,92] and related with lipid oxidation [29,75,83], green bean and butter flavors [66]. According to Fruet et al. [94], feeding with grass provided animals with significantly ($p < 0.001$) lower concentrations of furan, 2-pentyl (6.41 AU × 10^4/g fresh meat compared to the 17.49 AU × 10^4/g fresh meat of lambs reared intensively). This fact could reveal that the grass-based diet has a higher content of α-tocopherol, since the formation of furan, 2-pentyl is negatively correlated with said antioxidant [24]. On the contrary, the rest of furans were found in a higher concentration in lambs reared in the extensive range, being significant ($p < 0.001$) in the case of furan, 2-ethyl- and furan, 2,3-dihydro-. Despite this, the total content of furans remained significantly ($p < 0.01$) higher in intensively-reared lambs (18.13 vs. 13.38 AU × 10^4/g fresh) due to their higher contribution of furan, 2-pentyl. Additionally, the furan group represented a percentage of 2.90 and 1.66% of the total volatile compounds found in intensive and extensive lambs, respectively (Figure 1). These fractions are not very high, since furans represent the sixth and seventh family in lambs reared extensively and intensively, respectively. However, their occurrence can be very important, since these compounds are potential contributors to the rancid aroma of meat [76].

Finally, in the present research three sulfur compounds were identified in both intensively-reared and extensively-reared lambs, which were significantly ($p < 0.01$) affected by the production system (Table 5). Specifically, intensively-reared lambs produced a significantly ($p < 0.01$) higher concentration for the total content of these substances (48.30 vs. 23.81 AU × 10^4/g fresh). In disagreement with these findings, several studies displayed that sulfur compounds were present at higher concentration in grass-feed animals compared to animals fed with concentrates [6,14]. However, the higher content in our research can be related to the amount of the carbon disulfide, since it turned out to be the only sulfurous compound found in high levels in lambs reared intensively (45.75 vs. 21.34 AU × 10^4/g fresh). Despite this difference, disulfide carbon was the most abundant sulfur compound detected in both production systems. This substance can be derived from the enzymatic proteolysis of sulfur-containing amino acids [95] and/or from dithiocarbamate fungicides employed in agriculture [96]. Disulfide carbon could be important in the aromatic profile of lamb as it has been found to contribute to the overall aroma of packed meat [95] and possess a pleasant, sweet or ether-like odor [48]. Furthermore, Karabagias [48] concluded that carbon disulfide could be considered as a typical volatile compound of raw lamb meat. Contrary, dimethyl sulfide and dimethyl sulfone have been detected in significantly ($p < 0.01$) higher amounts in extensively-reared animals. These compounds are important because they can create adverse flavors in extensively-reared lambs. In this respect, dimethyl sulfone has been associated with unfavorable sensory descriptors [6]. Regarding the contribution of sulfur compounds on the volatile profile, this family represented 7.48 and 2.95% of the total volatile compounds in the lambs reared intensively and extensively, respectively. This presence can be considered important since, in addition to being the third and fifth most abundant family in intensively-fed and extensively-fed lambs, sulfur compounds contribute to the general aroma of meat [95].

A further consideration on the overall meat aroma profile is that lambs fed extensively displayed a significantly ($p < 0.001$) higher concentration of total volatile compounds (806.13 vs. 645.13 AU × 10^4/g fresh meat). This difference could suggest that extensively-reared lambs may have a higher flavor intensity than intensively-reared lambs. Furthermore, these outcomes are in line with those encountered by other studies, which found a greater flavor intensity in the meat of animals fed with pasture [15,18] or whose mothers were grazing at a pasture [18,97] in comparison to meat from animals fed with concentrates. This occurrence could be due to the fact that extensively-reared lambs contain a significantly ($p < 0.001$) higher fat content than intensively-reared lambs (Table 1), which can generate a greater amount of volatile compounds. In our study no significant ($p > 0.05$) correlations were found between intramuscular fat and total volatile content for either of the two production systems analyzed (r = 0.030; $p > 0.05$, for intensively-reared lambs; and r = 0.127; $p > 0.05$, for extensively-reared lambs). Similarly, the correlations between intramuscular fat and the different families of volatile compounds were low (r < 0.450) and not significant ($p > 0.05$) in any case (except for the correlation found in extensively-reared lambs for total ethers, where a significant correlation was observed (r = 0.634; $p < 0.05$)). However, when the individual volatile compounds were analyzed, a significant correlation (both positive and negative, depending on the volatile substance) between IMF and 180 compounds was detected. Among these compounds, it is important to highlight that strong, significant and positive correlations (r > 0.6 and $p < 0.01$) between IMF and the most important lipid-derived compounds, such as 1-propanol (r = 0.708; $p < 0.01$), 1-pentanol (r = 0.642; $p < 0.01$), 1-hexanol (r = 0.633; $p < 0.01$), nonanal (r = 0.639; $p < 0.01$), 2-butanone (r = 0.654; $p < 0.01$), 2-pentanone (r = 0.645; $p < 0.01$) and 2-heptanone (r = 0.684; $p < 0.01$) was observed. This fact confirms that the influence of the production system on the lipid content and that the lipid composition could likely be one of the main important factors for the release of characteristic volatile compounds that mainly consists of lipid-derived compounds, which, generally speaking, have a high impact on meat aroma due to their low odor thresholds.

4. Conclusions

The use of two different production systems (intensive and extensive) significantly affected the proximate composition of the BEDM lamb meat. Animals reared in an extensive production system presented the highest values of IMF and protein, while it demonstrated the lowest values for moisture and ash. In the same manner, the total concentration of volatile compounds was also affected, being higher in lambs reared in extensive regime, although it did not seem related to intramuscular fat content. However, the intramuscular fat content had a strong effect on the most individual volatile content derived from lipid reactions (lipolysis, lipid oxidation, etc.). Furthermore, most of the individual volatile compounds were also influenced by the production system, which could be related to both, and specific compounds linked to the diet or the variation of lipid fraction (lipid content and fatty acids), which highly influenced the release of lipid-derived volatile compounds.

Supplementary Materials: The following are available online at https://www.mdpi.com/article/10.3390/foods10071450/s1, Table S1: Chemical composition, ingredients and amounts of mineral and vitamin mix used in the diet of intensively-reared lambs.

Author Contributions: Conceptualization, V.A.P.C. and J.M.L.; formal analysis, N.E., R.B. and L.P.; writing—original draft preparation, N.E. and R.D.; writing—review and editing, R.D., U.G.-B., E.H. and J.M.L. All authors have read and agreed to the published version of the manuscript.

Funding: This research was funded by the EU ERA-NET programme (project "EcoLamb–Holistic Production to Reduce the Ecological Footprint of Meat"), grant number SusAn/0002/2016 through the Portuguese Foundation for Science and Technology (FCT) and the Agencia Estatal de Investigación (Acciones de Programación Conjunta Internacional) grant number PCIN-2017-053.

Institutional Review Board Statement: Ethical review and approval were waived for this study due to the compliance with national and/or European Regulations on animal husbandry and slaughter.

Informed Consent Statement: Not applicable.

Data Availability Statement: No new data were created or analyzed in this study. Data sharing is not applicable to this article.

Acknowledgments: Noemí Echegaray acknowledges Consellería de Cultura, Educación e Ordenación Universitaria (Xunta de Galicia) for granting with a predoctoral scholarship (Grant number IN606A-2018/002). Rubén Domínguez, Laura Purriños, Roberto Bermúdez and José M. Lorenzo are members of the HealthyMeat network and funded by CYTED (ref. 119RT0568). CIMO authors are grateful to FCT and FEDER under Programme PT2020 for the financial support to CIMO

(UIDB/00690/2020). Gonzales-Barron acknowledges the national funding by FCT, P.I., through the Institutional Scientific Employment Programme contract.

Conflicts of Interest: The authors declare no conflict of interest.

References

1. Font-i-Furnols, M.; Guerrero, L. Consumer preference, behavior and perception about meat and meat products: An overview. *Meat Sci.* **2014**, *98*, 361–371. [CrossRef]
2. Glitsch, K. Consumer perceptions of fresh meat quality: Cross-national comparison. *Br. Food J.* **2000**, *102*, 177–194. [CrossRef]
3. Watkins, P.J.; Frank, D.; Singh, T.K.; Young, O.A.; Warner, R.D. Sheepmeat flavor and the effect of different feeding systems: A review. *J. Agric. Food Chem.* **2013**, *61*, 3561–3579. [CrossRef] [PubMed]
4. Young, O.A.; Berdagué, J.L.; Viallon, C.; Rousset-Akrim, S.; Theriez, M. Fat-borne volatiles and sheepmeat odour. *Meat Sci.* **1997**, *45*, 183–200. [CrossRef]
5. Almela, E.; Jordán, M.J.; Martínez, C.; Sotomayor, J.A.; Bedia, M.; Bañón, S. Ewe's diet (pasture vs grain-based feed) affects volatile profile of cooked meat from light lamb. *J. Agric. Food Chem.* **2010**, *58*, 9641–9646. [CrossRef] [PubMed]
6. Young, O.A.; Lane, G.A.; Priolo, A.; Fraser, K. Pastoral and species flavour in lambs raised on pasture, lucerne or maize. *J. Sci. Food Agric.* **2003**, *83*, 93–104. [CrossRef]
7. Sañudo, C.; Alfonso, M.; Sanchez, A.; Berge, P.; Dransfield, E.; Zygoyiannis, D.; Stamataris, C.; Thorkelsson, G.; Valdimarsdottir, T.; Piasentier, E.; et al. Meat texture of lambs from different European production systems. *Aust. J. Agric. Res.* **2003**, *54*, 551–560. [CrossRef]
8. De-Arriba, R.; Sánchez-Andrés, A. Production and Productivity in Eastern and Western European Sheep Farming: A Comparative Analysis. Available online: http://www.lrrd.org/lrrd26/4/arri26066.htm (accessed on 3 July 2020).
9. Cadavez, V.A.P.; Popova, T.; Bermúdez, R.; Osoro, K.; Purriños, L.; Bodas, R.; Lorenzo, J.M.; Gonzales-Barron, U. Compositional attributes and fatty acid profile of lamb meat from Iberian local breeds. *Small Rumin. Res.* **2020**, *193*, 106244. [CrossRef]
10. Ruano, Z.M.; Cortinhas, A.; Carolino, N.; Gomes, J.; Costa, M.; Mateus, T.L. Gastrointestinal parasites as a possible threat to an endangered autochthonous Portuguese sheep breed. *J. Helminthol.* **2019**, *94*, e103. [CrossRef]
11. Mendelsohn, R. The challenge of conserving indigenous domesticated animals. *Ecol. Econ.* **2003**, *45*, 501–510. [CrossRef]
12. Paim, T.d.P.; Da Silva, A.F.; Martins, R.F.S.; Borges, B.O.; Lima, P.d.M.T.; Cardoso, C.C.; Esteves, G.I.F.; Louvandini, H.; McManus, C. Performance, survivability and carcass traits of crossbred lambs from five paternal breeds with local hair breed Santa Inês ewes. *Small Rumin. Res.* **2013**, *112*, 28–34. [CrossRef]
13. Cruz, B.C.; Cerqueira, J.; Araújo, J.P.; Gonzales-Barron, U.; Cadavez, V. Study of growth performance of Churra-Galega-Bragançana and Bordaleira-de-Entre-Douro-e-Minho lamb breeds. In Proceedings of the XVIII Jornadas sobre Producción Animal, Zaragoza, Spain, 7–8 May 2019; pp. 66–68.
14. Vasta, V.; Priolo, A. Ruminant fat volatiles as affected by diet. A review. *Meat Sci.* **2006**, *73*, 218–228. [CrossRef] [PubMed]
15. Priolo, A.; Micol, D.; Agabriel, J. Effects of grass feeding systems on ruminant meat colour and flavour. A review. *Anim. Res.* **2001**, *50*, 185–200. [CrossRef]
16. Vasta, V.; Ratel, J.; Engel, E. Mass spectrometry analysis of volatile compounds in raw meat for the authentication of the feeding background of farm animals. *J. Agric. Food Chem.* **2007**, *55*, 4630–4639. [CrossRef]
17. Sivadier, G.; Ratel, J.; Engel, E. Latency and persistence of diet volatile biomarkers in lamb fats. *J. Agric. Food Chem.* **2009**, *57*, 645–652. [CrossRef]
18. Vasta, V.; D'Alessandro, A.G.; Priolo, A.; Petrotos, K.; Martemucci, G. Volatile compound profile of ewe's milk and meat of their suckling lambs in relation to pasture vs. indoor feeding system. *Small Rumin. Res.* **2012**, *105*, 16–21. [CrossRef]
19. Naczk, M.; Shahidi, F. Extraction and analysis of phenolics in food. *J. Chromatogr. A* **2004**, *1054*, 95–111. [CrossRef]
20. Sebastiàn, I.; Viallon, C.; Berge, P. Analysis of the volatile fraction and the flavour characteristics of lamb: Relationships with the type of feeding. *Sci. Aliment.* **2003**, *23*, 497–511. [CrossRef]
21. Suzuky, J.; Bailey, M.E. Direct sampling capillary GLC analysis of flavor volatiles from ovine fat. *J. Agric. Food Chem.* **1985**, *33*, 343–347. [CrossRef]
22. Vlaeminck, B.; Fievez, V.; Van Laar, H.; Demeyer, D. Rumen odd and branched chain fatty acids in relation to in vitro rumen volatile fatty acid productions and dietary characteristics of incubated substrates. *J. Anim. Physiol. Anim. Nutr.* **2004**, *88*, 401–411. [CrossRef] [PubMed]
23. Raes, K.; Balcaen, A.; Dirinck, P.; De Winne, A.; Claeys, E.; Demeyer, D.; De Smet, S. Meat quality, fatty acid composition and flavour analysis in belgian retail beef. *Meat Sci.* **2003**, *65*, 1237–1246. [CrossRef]
24. Vasta, V.; Luciano, G.; Dimauro, C.; Röhrle, F.; Priolo, A.; Monahan, F.J.; Moloney, A.P. The volatile profile of longissimus dorsi muscle of heifers fed pasture, pasture silage or cereal concentrate: Implication for dietary discrimination. *Meat Sci.* **2011**, *87*, 282–289. [CrossRef]
25. ISO (International Organization for Standardization). Determination of moisture content, ISO 1442:1997 standard. In *International Standards Meat and Meat Products*; International Organization for Standardization: Genève, Switzerland, 1997.
26. ISO (International Organization for Standardization). Determination of nitrogen content, ISO 937:1978 standard. In *International Standards Meat and Meat Products*; International Organization for Standardization: Genève, Switzerland, 1978.

27. ISO (International Organization for Standardization). Determination of ash content, ISO 936:1998 standard. In *International Standards Meat and Meat Products*; International Organization for Standardization: Genève, Switzerland, 1998.
28. AOCS. *AOCS Official Procedure Am5-04. Rapid Determination of Oil/Fat Utilizing High Temperature Solvent Extraction*; American Oil Chemists Society: Urbana, IL, USA, 2005.
29. Domínguez, R.; Purriños, L.; Pérez-Santaescolástica, C.; Pateiro, M.; Barba, F.J.; Tomasevic, I.; Campagnol, P.C.B.; Lorenzo, J.M. Characterization of Volatile Compounds of Dry-Cured Meat Products Using HS-SPME-GC/MS Technique. *Food Anal. Methods* **2019**, *12*, 1263–1284. [CrossRef]
30. Gonzales-Barron, U.; Popova, T.; Bermúdez Piedra, R.; Tolsdorf, A.; Geß, H.; Pires, J.; Domínguez, R.; Chiesa, F.; Brugiapaglia, A.; Viola, I.; et al. Fatty acid composition of lamb meat from Italian and German local breeds. *Small Rumin. Res.* **2021**, *200*, 106384. [CrossRef]
31. Polidori, P.; Pucciarelli, S.; Cammertoni, N.; Polzonetti, V.; Vincenzetti, S. The effects of slaughter age on carcass and meat quality of Fabrianese lambs. *Small Rumin. Res.* **2017**, *155*, 12–15. [CrossRef]
32. Cividini, A.; Levart, A.; Žgur, S.; Kompan, D. Fatty acid composition of lamb meat from the autochthonous Jezersko-Solčava breed reared in different production systems. *Meat Sci.* **2014**, *97*, 480–485. [CrossRef] [PubMed]
33. Velasco, S.; Cañeque, V.; Pérez, C.; Lauzurica, S.; Díaz, M.T.; Huidobro, F.; Manzanares, C.; González, J. Fatty acid composition of adipose depots of suckling lambs raised under different production systems. *Meat Sci.* **2001**, *59*, 325–333. [CrossRef]
34. Gkarane, V.; Brunton, N.P.; Allen, P.; Gravador, R.S.; Claffey, N.A.; Diskin, M.G.; Fahey, A.G.; Farmer, L.J.; Moloney, A.P.; Alcalde, M.J.; et al. Effect of finishing diet and duration on the sensory quality and volatile profile of lamb meat. *Food Res. Int.* **2019**, *115*, 54–64. [CrossRef]
35. Aurousseau, B.; Bauchart, D.; Faure, X.; Galot, A.L.; Prache, S.; Micol, D.; Priolo, A. Indoor fattening of lambs raised on pasture. Part 1: Influence of stall finishing duration on lipid classes and fatty acids in the longissimus thoracis muscle. *Meat Sci.* **2007**, *76*, 241–252. [CrossRef]
36. Joy, M.; Ripoll, G.; Delfa, R. Effects of feeding system on carcass and non-carcass composition of Churra Tensina light lambs. *Small Rumin. Res.* **2008**, *78*, 123–133. [CrossRef]
37. Wilches, D.; Rovira, J.; Jaime, I.; Palacios, C.; Lurueña-Martínez, M.A.; Vivar-Quintana, A.M.; Revilla, I. Evaluation of the effect of a maternal rearing system on the odour profile of meat from suckling lamb. *Meat Sci.* **2011**, *88*, 415–423. [CrossRef]
38. Belhaj, K.; Mansouri, F.; Sindic, M.; Fauconnier, M.-L.; Boukharta, M.; Serghini Caid, H.; Elamrani, A. Effect of Rearing Season on Meat and Intramuscular Fat Quality of Beni-Guil Sheep. *J. Food Qual.* **2021**, *2021*, 1–9. [CrossRef]
39. Hajji, H.; Joy, M.; Ripoll, G.; Smeti, S.; Mekki, I.; Gahete, F.M.; Mahouachi, M.; Atti, N. Meat physicochemical properties, fatty acid profile, lipid oxidation and sensory characteristics from three North African lamb breeds, as influenced by concentrate or pasture finishing diets. *J. Food Compos. Anal.* **2016**, *48*, 102–110. [CrossRef]
40. Majdoub-Mathlouthi, L.; Saïd, B.; Say, A.; Kraiem, K. Effect of concentrate level and slaughter body weight on growth performances, carcass traits and meat quality of Barbarine lambs fed oat hay based diet. *Meat Sci.* **2013**, *93*, 557–563. [CrossRef]
41. Cañeque, V.; Velasco, S.; Díaz, M.; Pérez, C.; Huidobro, F.; Lauzurica, S.; Manzanares, C.; González, J. Effect of weaning age and slaughter weight on carcass and meat quality of Talaverana breed lambs raised at pasture. *Anim. Sci.* **2001**, *73*, 85–95. [CrossRef]
42. Popova, T.; Gonzales-Barron, U.; Cadavez, V. A meta-analysis of the effect of pasture access on the lipid content and fatty acid composition of lamb meat. *Food Res. Int.* **2015**, *77*, 476–483. [CrossRef]
43. Ekiz, B.; Demirel, G.; Yilmaz, A.; Ozcan, M.; Yalcintan, H.; Kocak, O.; Altinel, A. Slaughter characteristics, carcass quality and fatty acid composition of lambs under four different production systems. *Small Rumin. Res.* **2013**, *114*, 26–34. [CrossRef]
44. Boughalmi, A.; Araba, A. Effect of feeding management from grass to concentrate feed on growth, carcass characteristics, meat quality and fatty acid profile of Timahdite lamb breed. *Small Rumin. Res.* **2016**, *144*, 158–163. [CrossRef]
45. Nuernberg, K.; Nuernberg, G.; Ender, K.; Dannenberger, D.; Schabbel, W.; Grumbach, S.; Zupp, W.; Steinhart, H. Effect of grass vs. concentrate feeding on the fatty acid profile of different fat depots in lambs. *Eur. J. Lipid Sci. Technol.* **2005**, *107*, 737–745. [CrossRef]
46. Ates, S.; Keles, G.; Demirci, U.; Dogan, S.; Kirbas, M.; Filley, S.J.; Parker, N.B. The effects of feeding system and breed on the performance and meat quality of weaned lambs. *Small Rumin. Res.* **2020**, *192*, 106225. [CrossRef]
47. Ye, Y.; Schreurs, N.M.; Johnson, P.L.; Corner-Thomas, R.A.; Agnew, M.P.; Silcock, P.; Eyres, G.T.; Maclennan, G.; Realini, C.E. Carcass characteristics and meat quality of commercial lambs reared in different forage systems. *Livest. Sci.* **2020**, *232*, 103908. [CrossRef]
48. Karabagias, I.K. Volatile profile of raw lamb meat stored at 4 ± 1 °C: The potential of specific aldehyde ratios as indicators of lamb meat quality. *Foods* **2018**, *7*, 40. [CrossRef] [PubMed]
49. Meynier, A.; Novelli, E.; Chizzolini, R.; Zanardi, E.; Gandemer, G. Volatile compounds of commercial Milano salami. *Meat Sci.* **1999**, *51*, 175–183. [CrossRef]
50. Sivadier, G.; Ratel, J.; Bouvier, F.; Engel, E. Authentication of meat products: Determination of animal feeding by parallel GC-MS analysis of three adipose tissues. *J. Agric. Food Chem.* **2008**, *56*, 9803–9812. [CrossRef] [PubMed]
51. Rios, J.J.; Fernández-García, E.; Mínguez-Mosquera, M.I.; Pérez-Gálvez, A. Description of volatile compounds generated by the degradation of carotenoids in paprika, tomato and marigold oleoresins. *Food Chem.* **2008**, *106*, 1145–1153. [CrossRef]
52. Binnie, J.; Cape, J.N.; Mackie, N.; Leith, I.D. Exchange of organic solvents between the atmosphere and grass—The use of open top chambers. *Sci. Total Environ.* **2002**, *285*, 53–67. [CrossRef]

53. Smith, K.E.C.; Jones, K.C. Particles and vegetation: Implications for the transfer of particle-bound organic contaminants to vegetation. *Sci. Total Environ.* **2000**, *246*, 207–236. [CrossRef]
54. Franke, E.N. Recent advances in the chemistry of rancidity of fats. In *Recent Advances in the Chemistry of Meat*; The Royal Society of Chemistry: London, UK, 1984; pp. 87–118.
55. Flores, M. Understanding the implications of current health trends on the aroma of wet and dry cured meat products. *Meat Sci.* **2018**, *144*, 53–61. [CrossRef]
56. Insausti, K.; Beriain, M.; Gorraiz, C.; Purroy, A. Volatile compounds of raw beef from 5 local spanish cattle breeds stored under modified atmosphere. *Sens. Nutr. Qual. Food* **2002**, *67*, 1580–1589. [CrossRef]
57. Mottram, D.S. Flavour formation in meat and meat products: A review. *Food Chem.* **1998**, *62*, 415–424. [CrossRef]
58. Ha, J.K.; Lindsay, R.C. Volatile alkylphenols and thiophenol in species-related characterizing flavors of red meats. *J. Food Sci.* **1991**, *56*, 1197–1202. [CrossRef]
59. Erasmus, S.W.; Muller, M.; Hoffman, L.C. Authentic sheep meat in the European Union: Factors influencing and validating its unique meat quality. *J. Sci. Food Agric.* **2017**, *97*, 1979–1996. [CrossRef]
60. Watkins, P.J.; Kearney, G.; Rose, G.; Allen, D.; Ball, A.J.; Pethick, D.W.; Warner, R.D. Effect of branched-chain fatty acids, 3-methylindole and 4-methylphenol on consumer sensory scores of grilled lamb meat. *Meat Sci.* **2014**, *96*, 1088–1094. [CrossRef] [PubMed]
61. Garcia, C.; Berdagué, J.J.; Antequera, T.; López-Bote, C.; Córdoba, J.J.; Ventanas, J. Volatile components of dry cured Iberian ham. *Food Chem.* **1991**, *41*, 23–32. [CrossRef]
62. Barbieri, G.; Bolzoni, L.; Parolari, G.; Virgili, R.; Buttini, R.; Careri, M.; Mangia, A. Flavor compounds of dry-cured ham. *J. Agric. Food Chem.* **1992**, *40*, 2389–2394. [CrossRef]
63. Domínguez, R.; Pateiro, M.; Gagaoua, M.; Barba, F.J.; Zhang, W.; Lorenzo, J.M. A comprehensive review on lipid oxidation in meat and meat products. *Antioxidants* **2019**, *8*, 429. [CrossRef]
64. Elmore, J.S.; Cooper, S.L.; Enser, M.; Mottram, D.S.; Sinclair, L.A.; Wilkinson, R.G.; Wood, J.D. Dietary manipulation of fatty acid composition in lamb meat and its effect on the volatile aroma compounds of grilled lamb. *Meat Sci.* **2005**, *69*, 233–242. [CrossRef]
65. Morand-Fehr, P.; Tran, G. La fraction lipidique des aliments et les corps gras utilisés en alimentation animale. *Prod. Anim.* **2001**, *14*, 285–302. [CrossRef]
66. Calkins, C.R.; Hodgen, J.M. A fresh look at meat flavor. *Meat Sci.* **2007**, *77*, 63–80. [CrossRef]
67. Lane, G.A.; Fraser, K. A comparison of phenol and indole flavour compounds in fat, and of phenols in urine of cattle fed pasture or grain. *N. Zeal. J. Agric. Res.* **1999**, *42*, 289–296. [CrossRef]
68. Gravador, R.S.; Serra, A.; Luciano, G.; Pennisi, P.; Vasta, V.; Mele, M.; Pauselli, M.; Priolo, A. Volatiles in raw and cooked meat from lambs fed olive cake and linseed. *Animal* **2015**, *9*, 715–722. [CrossRef] [PubMed]
69. Coppa, M.; Martin, B.; Pradel, P.; Leotta, B.; Priolo, A.; Vasta, V. Milk volatile compounds to trace cows fed a hay-based diet or different grazing systems on upland pastures. *J. Agric. Food Chem.* **2011**, *59*, 4947–4954. [CrossRef] [PubMed]
70. Selli, S.; Cayhan, G.G. Analysis of volatile compounds of wild gilthead sea bream (Sparus aurata) by simultaneous distillation-extraction (SDE) and GC-MS. *Microchem. J.* **2009**, *93*, 232–235. [CrossRef]
71. Argemí-Armengol, I.; Villalba, D.; Tor, M.; Pérez-Santaescolástica, C.; Purriños, L.; Lorenzo, J.M.; Álvarez-Rodríguez, J. The extent to which genetics and lean grade affect fatty acid profiles and volatile compounds in organic pork. *PeerJ* **2019**, *7*, e7322. [CrossRef]
72. Maggiolino, A.; Lorenzo, J.M.; Marino, R.; della Malva, A.; Centoducati, P.; De Palo, P. Foal meat volatile compounds: Effect of vacuum ageing on semimembranosus muscle. *J. Sci. Food Agric.* **2019**, *99*, 1660–1667. [CrossRef] [PubMed]
73. Van, H.; Hwang, I.; Jeong, D.; Touseef, A. Principle of Meat Aroma Flavors and Future Prospect. In *Latest Research into Quality Control*; Akyar, I., Ed.; IntechOpen: London, UK, 2012; pp. 145–176.
74. Frankel, E.N. *Lipid Oxidation*, 2nd ed.; Oily Press: Dundee, Scotland, 1998.
75. DeMan, J.H. Lipids. In *Principles of Food Chemistry*; Reinhold, V.N., Ed.; Springer: New York, NY, USA, 1990.
76. Ortuño, J.; Serrano, R.; Bañón, S. Use of dietary rosemary diterpenes to inhibit rancid volatiles in lamb meat packed under protective atmosphere. *Animal* **2016**, *10*, 1391–1401. [CrossRef]
77. Marco, A.; Navarro, J.L.; Flores, M. Quantitation of selected odor-active constituents in dry fermented sausages prepared with different curing salts. *J. Agric. Food Chem.* **2007**, *55*, 3058–3065. [CrossRef]
78. Larick, D.K.; Tuener, B.E. Headspace volatiles and sensory characteristics of ground beef from forage- and grain-fed heifers. *J. Food Sci.* **1990**, *55*, 649–654. [CrossRef]
79. Du, X. Determination of flavor substances in fermented pork by GC-MS. *Meat Res.* **2012**, *26*, 34–36.
80. Elmore, J.S.S.; Mottram, D.S.D.S.; Enser, M.; Wood, J.D.J.D. Effect of the polyunsaturated fatty acid composition of beef muscle on the profile of aroma volatiles. *J. Agric. Food Chem.* **1999**, *47*, 1619–1625. [CrossRef] [PubMed]
81. Resconi, V.C.; Campo, M.M.; Montossi, F.; Ferreira, V.; Sañudo, C.; Escudero, A. Relationship between odour-active compounds and flavour perception in meat from lambs fed different diets. *Meat Sci.* **2010**, *85*, 700–706. [CrossRef] [PubMed]
82. Caporaso, F.; Sink, J.D.; Dimick, P.S.; Mussinan, C.J.; Sanderson, A. Volatile flavor constituents of ovine adipose tissue. *J. Agric. Food Chem.* **1977**, *25*, 1230–1234. [CrossRef]
83. Gkarane, V.; Brunton, N.P.; Harrison, S.M.; Gravador, R.S.; Allen, P.; Claffey, N.A.; Diskin, M.G.; Fahey, A.G.; Farmer, L.J.; Moloney, A.P.; et al. Volatile profile of grilled lamb as affected by castration and age at slaughter in two breeds. *J. Food Sci.* **2018**, *83*, 2466–2477. [CrossRef]

84. Manner, W.; Maxwell, R.J.; Williams, J.E. Effects of dietary regimen and tissue site on bovine fatty acid profiles. *J. Anim. Sci.* **1984**, *59*, 109–121. [CrossRef]
85. Gargouri, M.; Drouet, P.; Legoy, M.D. Synthesis of a novel macrolactone by lipase-catalyzed intra-esterification of hydroxy-fatty acid in organic media. *J. Biotechnol.* **2002**, *92*, 259–266. [CrossRef]
86. Urbach, G. Effect of feed on flavor in dairy foods. *J. Dairy Sci.* **1990**, *73*, 3639–3650. [CrossRef]
87. Krvavica, M.; Bradaš, M.; Rogošić, J.; Jug, T.; Vnučec, I.; Marušić Radovčić, N. Volatile aroma compounds of Lika lamb. *MESO* **2015**, *3*, 469–476.
88. Osorio, M.T.; Zumalacárregui, J.M.; Cabeza, E.A.; Figueira, A.; Mateo, J. Effect of rearing system on some meat quality traits and volatile compounds of suckling lamb meat. *Small Rumin. Res.* **2008**, *78*, 1–12. [CrossRef]
89. Petričević, S.; Marušić Radovčić, N.; Lukić, K.; Listeš, E.; Medić, H. Differentiation of dry-cured hams from different processing methods by means of volatile compounds, physico-chemical and sensory analysis. *Meat Sci.* **2018**, *137*, 217–227. [CrossRef]
90. Vasta, V.; Jerónimo, E.; Brogna, D.M.R.; Dentinho, M.T.P.; Biondi, L.; Santos-Silva, J.; Priolo, A.; Bessa, R.J.B. The effect of grape seed extract or Cistus ladanifer L. on muscle volatile compounds of lambs fed dehydrated lucerne supplemented with oil. *Food Chem.* **2010**, *119*, 1339–1345. [CrossRef]
91. Zhang, C.; Zhang, H.; Liu, M.; Zhao, X.; Luo, H. Effect of breed on the volatile compound precursors and odor profile attributes of lamb meat. *Foods* **2020**, *9*, 1178. [CrossRef]
92. Del Bianco, S.; Natalello, A.; Luciano, G.; Valenti, B.; Monahan, F.; Gkarane, V.; Rapisarda, T.; Carpino, S.; Piasentier, E. Influence of dietary cardoon meal on volatile compounds and flavour in lamb meat. *Meat Sci.* **2020**, *163*, 108086. [CrossRef]
93. Bueno, M.; Resconi, V.C.; Campo, M.M.; Cacho, J.; Ferreira, V.; Escudero, A. Gas chromatographic-olfactometric characterisation of headspace and mouthspace key aroma compounds in fresh and frozen lamb meat. *Food Chem.* **2011**, *129*, 1909–1918. [CrossRef]
94. Fruet, A.P.B.; Trombetta, F.; Stefanello, F.S.; Speroni, C.S.; Donadel, J.Z.; De Souza, A.N.M.; Rosado Júnior, A.; Tonetto, C.J.; Wagner, R.; De Mello, A.; et al. Effects of feeding legume-grass pasture and different concentrate levels on fatty acid profile, volatile compounds, and off-flavor of the M. longissimus thoracis. *Meat Sci.* **2018**, *140*, 112–118. [CrossRef] [PubMed]
95. Saraiva, C.; Oliveira, I.; Silva, J.A.; Martins, C.; Ventanas, J.; García, C. Implementation of multivariate techniques for the selection of volatile compounds as indicators of sensory quality of raw beef. *J. Food Sci. Technol.* **2015**, *52*, 3887–3898. [CrossRef] [PubMed]
96. Kontou, S.; Tsipi, D.; Tzia, C. Stability of the dithiocarbamate pesticide maneb in tomato homogenates during cold storage and thermal processing. *Food Addit. Contam.* **2004**, *21*, 1083–1089. [CrossRef] [PubMed]
97. Maiorano, G.; Kowaliszyn, B.; Martemucci, G.; Breeding, G.A.; Amendola, G. The effect of production system information on consumer expectation and acceptability of Leccese lamb meat. *Ann. Food Sci. Technol.* **2010**, *11*, 9–13.

Article

Comprehensive SPME-GC-MS Analysis of VOC Profiles Obtained Following High-Temperature Heating of Pork Back Fat with Varying Boar Taint Intensities

Clément Burgeon [1,*], Alice Markey [1], Marc Debliquy [2], Driss Lahem [3], Justine Rodriguez [2], Ahmadou Ly [3] and Marie-Laure Fauconnier [1]

[1] Laboratory of Chemistry of Natural Molecules, Gembloux Agro-Bio Tech, Université de Liège, Passage des Déportés 2, 5030 Gembloux, Belgium; Alice.Markey@student.uliege.be (A.M.); marie-laure.fauconnier@uliege.be (M.-L.F.)
[2] Service de Science des Matériaux, Faculté Polytechnique, Université de Mons, Rue de l'Epargne 56, 7000 Mons, Belgium; marc.debliquy@umons.ac.be (M.D.); Justine.RODRIGUEZ@umons.ac.be (J.R.)
[3] Materia Nova ASBL, Materials R&D Centre, Parc Initialis, Avenue Nicolas Copernic 3, 7000 Mons, Belgium; driss.lahem@materianova.be (D.L.); ahmadou.ly@materianova.be (A.L.)
* Correspondence: cburgeon@uliege.be

Citation: Burgeon, C.; Markey, A.; Debliquy, M.; Lahem, D.; Rodriguez, J.; Ly, A.; Fauconnier, M.-L. Comprehensive SPME-GC-MS Analysis of VOC Profiles Obtained Following High-Temperature Heating of Pork Back Fat with Varying Boar Taint Intensities. *Foods* **2021**, *10*, 1311. https://doi.org/10.3390/foods10061311

Academic Editor: Paulo Eduardo Sichetti Munekata

Received: 21 May 2021
Accepted: 1 June 2021
Published: 7 June 2021

Publisher's Note: MDPI stays neutral with regard to jurisdictional claims in published maps and institutional affiliations.

Copyright: © 2021 by the authors. Licensee MDPI, Basel, Switzerland. This article is an open access article distributed under the terms and conditions of the Creative Commons Attribution (CC BY) license (https://creativecommons.org/licenses/by/4.0/).

Abstract: Boar taint detection is a major concern for the pork industry. Currently, this taint is mainly detected through a sensory evaluation. However, little is known about the entire volatile organic compounds (VOCs) profile perceived by the assessor. Additionally, many research groups are working on the development of new rapid and reliable detection methods, which include the VOCs sensor-based methods. The latter are susceptible to sensor poisoning by interfering molecules produced during high-temperature heating of fat. Analyzing the VOC profiles obtained by solid phase microextraction gas chromatography–mass spectrometry (SPME-GC-MS) after incubation at 150 and 180 °C helps in the comprehension of the environment in which boar taint is perceived. Many similarities were observed between these temperatures; both profiles were rich in carboxylic acids and aldehydes. Through a principal component analysis (PCA) and analyses of variance (ANOVAs), differences were highlighted. Aldehydes such as (*E*,*E*)-nona-2,4-dienal exhibited higher concentrations at 150 °C, while heating at 180 °C resulted in significantly higher concentrations in fatty acids, several amide derivatives, and squalene. These differences stress the need for standardized parameters for sensory evaluation. Lastly, skatole and androstenone, the main compounds involved in boar taint, were perceived in the headspace at these temperatures but remained low (below 1 ppm). Higher temperature should be investigated to increase headspace concentrations provided that rigorous analyses of total VOC profiles are performed.

Keywords: back fat; boar taint; entire male pig; GC-MS; lipid oxidation; meat quality; pork meat; SPME; VOC

1. Introduction

Nowadays, a top priority for the pork industry is being able to correctly discriminate tainted from untainted boar carcasses. In fact, boar taint is a strong and unpleasant smell found in the meat of some uncastrated male pigs. This smell appears upon cooking of boar tainted meat and is due to the release of a complex mixture of molecules. The major molecules responsible for this smell are the steroid androstenone (5-α-androst-16-en-3-one) and the tryptophan metabolite skatole (3-methylindole), which are well-known for their urine and fecal smell, respectively [1,2].

To prevent the development of such molecules, surgical castration without pain relief has often been used worldwide given that it is a fast, cheap, and handy castration technique for farmers. However, this practice has been criticized for the pain and stress that it inflicts to piglets. Hence, alternatives to surgical castration have been suggested and are now being promoted [3]. Out of all, two castration techniques appear more realistic: immunocastration (i.e., testicular functions are deactivated through the neutralization of the hypothalamic–pituitary–gonadal axis hormones [4]) and rearing of entire males. Whether it is to discriminate tainted uncastrated male pigs or simply to ensure that immunocastration has functioned correctly, the detection of tainted carcasses is an essential step in the slaughtering process.

Currently, many research studies are taking place to develop new detection methods that are ideally low cost (less than 1.30 euro/analysis), fast (less than 10 s/analysis), 100% specific and sensitive (no false negatives and no false positives), and automated [5]. These criteria are essential for methods to be used on the slaughter line.

Several detection principles have been investigated throughout the years [6]. Mass spectrometry-based methods have recently been examined and have shown interesting results. Rapid evaporative ionization mass spectrometry (REIMS) provided highly accurate classification of tainted and untainted samples at a fast speed and has shown its potential to be used for online applications given its hand-held sampling tool and estimated low cost [7]. Laser diode thermal desorption–tandem mass spectrometry (LDTD-MS/MS) has also been thoroughly investigated [8–10]. This method achieved good validation criteria, fast analysis (once sample preparation has been performed, analysis in itself takes less than 10 s/sample), and is currently being tested in a Danish slaughterhouse [11]. However, both methods would require substantial investment (expensive instruments and need for skilled staff), which could lead to reluctance in their application.

Other methods recently tested and presenting lower investment cost are devices based on Raman spectroscopy [12,13] and a new specific sensor system based on screen-printed carbon electrodes [14,15]. Additionally, these techniques are easy to use given the hand-held measuring tool. However, both still need further validation given high prediction errors for Raman spectroscopy and the absence of real slaughterhouse testing with the sensor system.

The rapid detection of boar taint through volatile organic compounds (VOCs) detection has also been widely studied. Some researchers have tried using gas chromatography mass spectrometry (GC-MS) for this purpose [16,17]. However, the high initial investment and long (i.e., minimum 3.5 min [16]) analysis time remained two main drawbacks of GC-MS methods.

Boar taint detection through the use of e-noses has been extensively studied some years ago [18–22]. An e-nose is composed of an array of sensors for which a response is induced when gases, and in this case VOCs, are perceived at their surface. In a recent review by Burgeon et al. (2021) [6], the great potential of new sensor material for skatole and androstenone has been discussed, and this review concluded that sensor-based methods might be a solution for the rapid slaughterhouse detection of boar taint provided that it is able to detect low headspace concentrations of skatole and androstenone in a VOCs-rich environment. This working environment is due to the extraction conditions used to volatilize skatole and androstenone.

In fact, skatole and androstenone are lipophilic molecules with low vapor pressure (7.3×10^{-4} kPa and 1.3×10^{-6} kPa at 25 °C, respectively); hence, fat must be heated at high temperatures to allow the volatilization of these molecules. This heating leads to the release of a variety of molecules. Most of these molecules are products of lipids degradation (oxidation of fatty acids starting at 70 °C [23]). Lipids can oxidize in three main ways: autoxidation, enzymatic-catalyzed oxidation, and photo-oxidation. However, the most probable oxidation mechanism during fat heating remains autoxidation where the unsaturated fatty acids react with oxygen, which is activated by temperature in this case, to produce free radicals. These free radicals are unstable and therefore decompose to form various molecules, including acids, alcohols, esters, ketones hydrocarbons, and aldehydes. The latter are present in significant quantities in products that underwent oxidation processes [24].

Such a VOCs-rich environment can quickly lead to sensor poisoning, i.e., binding of VOCs to the sensor's surface, and in turn lead to temporal sensor drift. Such drift is defined as the gradual deviation of the sensor's response when exposed to the same molecule in the same environment [25]. Understanding the VOCs environment in which the volatilization of skatole and androstenone takes place is primordial, as this could help in creating new drift-reduction solutions, which are physical solutions (such as filters) aiming to reduce interfering VOCs present in the headspace but also creating more robust drift correction models taking such environments into account.

Until now, none of the above-mentioned methods have stood out compared to the others, and that is why current slaughterhouse boar taint detection is still performed either through a colorimetric method [26] or mainly by sensory evaluation [27].

Hence, the objective of this research was to examine elevated temperature VOC profiles to facilitate new sensor development, gain the understanding of VOCs perceived during boar taint sensory evaluations, and lastly help in understanding which VOCs perceived by the consumers during the cooking of pork meat are lipid-derived. The 150 and 180 °C temperatures were used in the current study, as they are frequently encountered for sensory evaluation in the frame of boar taint detection [28–32] and appear in the range of temperatures used for cooking by consumers [33].

Rius et al. (2005) [34] have already analyzed VOCs produced when heating fat at a temperature of 120 °C. However, only back fat with low concentrations in skatole and androstenone was analyzed, and comparisons of heating temperatures were not performed.

To the best of our knowledge, our study is the first providing a thorough understanding and comparisons of VOC profiles obtained following the heating of sow fat as well as tainted and untainted boar fat at two elevated temperatures (150 and 180 °C) and sampling and analysis by solid phase microextraction gas chromatography–mass spectrometry (SPME-GC-MS).

2. Materials and Methods

2.1. Samples

Sow back fat ($n = 6$), tainted ($n = 7$) and untainted boar fat ($n = 7$) were collected from a local slaughterhouse. Sow fat was randomly selected. Tainted and untainted boar fat, on the other hand, were selected after these had been checked for boar taint by a trained assessor through an online human nose detection method (soldering iron). The collected samples were frozen at $-20\ °C$ at the slaughterhouse, transported in a cooler, and stored again at $-20\ °C$. The presence or absence of boar taint was confirmed through the quantification of skatole and androstenone in fat by high-performance liquid chromatography fluorescence detection (HPLC-FD), which is described later in this section.

2.2. Chemicals

Methanol (CAS n° 67-56-1, HPLC grade, Sigma-Aldrich, Darmstadt, Germany), dansylhydrazine (CAS n° 33008-06-9, Sigma-Aldrich, Darmstadt, Germany), boron trifluoride (BF_3) at 20% in methanol v/v (CAS n° 373-57-9, VWR, Darmstadt, Germany), phosphoric acid (H_3PO_4) (CAS n° 7664-38-2, Sigma-Aldrich, Darmstadt, Germany), acetonitrile (CAS n° 75-05-8, HPLC grade, Supelco, Darmstadt, Germany), tetrahydrofuran (CAS n° 109-99-9, HPLC grade, Supelco, Darmstadt, Germany), liquid nitrogen (CAS n° 7727-37-9, Nippon Gases, Schoten, Belgium), 2,3-dimethylindole (CAS n°91-55-4, Sigma Aldrich, Darmstadt, Germany), skatole (CAS n° 83-34-1, Sigma Aldrich, Darmstadt, Germany), and androstenone (CAS n° 18339-16-7, Sigma Aldrich, Darmstadt, Germany) were used in this experiment.

2.3. Skatole and Androstenone Quantification in Back Fat

This analysis allowed quantifying the skatole and androstenone content in both tainted and untainted boar fat samples. Boar fat is considered tainted if skatole concentrations are above the thresholds of 200 ng g^{-1} of fat and/or above 1000 ng g^{-1} for androstenone.

These thresholds were selected given that the commonly accepted threshold generally range from 200 to 250 ng g^{-1} of fat for skatole and 500 to 1000 ng g^{-1} for androstenone [35]. Quantification of these molecules in back fat was performed on the basis of a method by Hansen-Moller (1994) [36], which consists of a methanolic extraction of the molecules, derivatization of androstenone, and analysis by high-performance liquid chromatography fluorescence detection (HPLC-FD). This protocol was slightly adapted as described in this section.

2.3.1. Extraction of Androstenone and Skatole

Two mL of methanol was added to 0.50 g of back fat cut into pieces (0.5 cm square). The sample was homogenized by an Ultra-Turrax T25 (Janke & Kunkel, Straufen, Germany) for 30 s at 13,500 rpm. Then, 500 µL of methanol was added, and the sample was homogenized again for 30 s with the Ultra-Turrax; finally, 500 µL of methanol was added and homogenized for 1 min with the Ultra-Turrax. The sample was ultrasonicated for 5 min and placed in an ice bath for 15 min before centrifugation at 7700 rpm at 4 °C. Then, the supernatant was passed through a 0.45 µm filter paper (Whatmann, Darmstadt, Germany), and 140 µL was put in vial for analysis.

2.3.2. Derivatization

The autosampler was programmed to mix 30 µL of 2% dansylhydrazine in methanol, 4.4 µL of water, and 10 µL of 20% v/v BF$_3$ with 140 µL of methanolic extract. A reaction time of 5 min was observed; then, 20 µL of the incubated sample was injected into HPLC.

2.3.3. High-Performance Liquid Chromatography Fluorescence Detection (HPLC-FD)

The analysis was performed by HPLC (1260 Infinity, Agilent Technologies, Santa Clara, CA, USA) with a kinetex column EVO C18 100 A (150 × 3.0 mm × 5 µm, Phenomenex, Utrecht, Belgium) and a precolumn AJO-9297, EVO C18 (Phenomenex, Utrecht, Belgium). The solutions for the mobile phase are prepared as follows: (A) H$_3$PO$_4$/deionized water (1:1000 v/v); (B) acetonitrile; (C) THF/deionized water (99:1 v/v). The elution gradient profile runs as presented in Table 1. The mobile phase was pumped at a flow rate of 0.5 mL min^{-1} throughout the process.

Table 1. Elution gradient for the separation of skatole and androstenone on an High-Performance Liquid Chromatography Fluorescence Detection (HPLC) system.

Time (min)	H3PO4/Deionized Water (1:1000 v/v)	Acetonitrile	THF/Deionized Water (99:1 v/v)
0	73	0	27
5.3	73	0	27
7.3	42	24	34
13	42	24	34
13.3	10	0	90
18	10	0	90
24	73	0	27
28	73	0	27

The detection with a fluorescence detector (FD) (Agilent Infinity 1260) was performed with an excitation wavelength of 285 nm and emission wavelength of 340 nm for skatole and 346 nm for excitation and 521 nm for emission of androstenone. The wavelength change takes place after 12 min of elution.

2.3.4. Quantification of Skatole and Androstenone

Quantification of skatole and androstenone was made possible with matrix-matched calibration curves. These were prepared with sow fat (very low concentrations in skatole and absence of androstenone) that had been previously spiked with standards solutions. Calibrations curves were prepared for concentrations ranging from 45 to 500 ng/g for skatole and from 240 to 5000 ng/g for androstenone.

2.4. Analysis of VOCs Found in the Headspace of Heated Back Fat Samples

VOC profiles were established following 6 different analyses (i.e., 6 modalities): heating of sow fat at 150 °C, untainted boar fat at 150 °C, tainted boar fat at 150 °C, sow fat at 180 °C, untainted boar fat at 180 °C, and tainted boar fat at 180 °C. The analyses were performed as described in this section.

2.4.1. Sample Preparation

First, 2.5 g of back fat was cut and then cooled by adding liquid nitrogen (-196 °C). The sample is ground for 5 s with an A11 basic IKA analytical grinder. Before recovering 1.0 g of sample in a vial, liquid nitrogen is added to the sample to freeze it. The sample is stored at -20 °C until analysis.

2.4.2. SPME-GC-MS VOCs Analysis

Before proceeding to the headspace solid phase microextraction GC-MS analysis (HS-SPME-GC-MS), 1 µL of 2,3-dimethylindole at 125 µg mL^{-1} in methanol is added on the inside of the 20 mL vial, which is immediately sealed with a magnetic screw cap with a PTFE septum (Sigma-Aldrich, Darmstadt, Germany).

Incubation of the sample takes place at 150 °C (for the first analysis) or 180 °C (for the second analysis) for 20 min in a heated agitator (Gerstel, Mülheim an der Ruhr, Germany). Then, sampling of VOCs was achieved with a divinylbenzene/carboxen/polydimethylsiloxane (DVB/CAR/PDMS, 50/30 µm) SPME fiber (Supelco, Darmstadt, Germany) through a 5 min exposition in the headspace. The vials were shaken at 250 rpm (agitator on/off time: 10 s/1 s) during incubation and extraction. Desorption of the extracted and captured VOCs takes place for 2 min. Injection was performed in splitless mode at 270 °C. The fiber was conditioned for 20 min at injection temperature. Analyses were performed by GC-MS (7890A-5975C, Agilent Technologies, Santa Clara, CA, USA) equipped with an HP-5 MS capillary column (30 m × 250 µm × 0.25 µm, Agilent Technologies, Santa Clara, CA, USA). Helium was used as a carrier gas at a flow rate of 1.2 mL/min. The oven temperature program was as follows: starting at 40 °C with a hold for 3 min; then, there is an increase of 5 °C/min up to 300 °C with a hold for 5 min. The mass spectrometer was set to have a temperature of 230 °C at the ion source and 150 °C at the quadrupole. The mass spectrometer was programmed with a SCAN/SIM acquisition mode. In SIM mode, the targeted ions were (quantitative ions in bold): 77, 103, and **130** for skatole; 130 and **144** for 2,3-dimethyl-indole; and lastly, 239, 257, and **272** for androstenone. The SIM mode allowed for semi-quantification of skatole and androstenone in the headspace using the following formula:

$$\text{Target (ppb)} = (\text{area}_{\text{target quant. ion}} / \text{area}_{\text{I.S. quant. ion}}) \times \text{mass of I.S.} \times (1/\text{vial volume}) \times \text{correction factor}. \quad (1)$$

The correction factor corresponds to 2.5 for skatole and 1/34 for androstenone; IS corresponds to the internal standard, i.e., 2,3-dimethylindole in this case.

In SCAN mode, mass spectra were scanned from 35 to 500 amu. Then, component identification was performed by comparison of the obtained spectra with mass spectra in a reference database (NIST17). Additionally, experimental retention indices (RI) were calculated following the injection of a mixture of n-alkanes C_8-C_{30} (Sigma Aldrich, Darmstadt, Germany) under the same chromatographic conditions as those previously mentioned. This allowed the comparison of these RI to literature RI. Lastly, pure standards were injected for skatole (CAS n° 83-34-1, Sigma Aldrich) and androstenone (CAS n° 18339-16-7, Sigma Aldrich, Darmstadt, Germany) to ensure identification [37–39].

2.5. Data Analysis

General VOC profiles were established through a chromatographic deconvolution process (Agilent MassHunter Unknowns Analysis), and chromatographic areas were obtained for each VOC. Then, these results were used in two different ways. In the first case, they were first reported as a percentage of the total chromatographic area to allow a general analysis. General linear models (GLMs) were performed on these data to validate some observations. Fat type was used as a fixed factor and incubation temperature was used as a covariate for GLM. In the second case, the chromatographic area data were mass-normalized, auto-scaled, and log-transformed (generalized logarithm transformation) to obtain a distribution of the variables closer to normal and make results more comparable. Then, a principal component analysis (PCA) as well as a heatmap were generated with these normalized data. One-way analyses of variance (ANOVAs) were performed on the normalized data of the top 25 contributors to the differences observed. The PCA and heatmap were carried out on metaboanalyst [40]. Pearson correlation coefficients were determined for the skatole and androstenone headspace and content concentrations data. These coefficients as well as GLMs mentioned earlier were established with the Minitab 19 software (Minitab Inc., State College, PA, USA). Headspace/content correlation plots for skatole and androstenone were performed on Excel (Microsoft Office 2016).

3. Results and Discussion

In this section, results concerning the analysis of VOC profiles obtained with the high incubation temperatures used, i.e., 150 and 180 °C, will first be examined. Fat can be heated even more for boar taint detection; however, lipid oxidation occurs at a greater extent in this case. Therefore, in this research, 150 and 180 °C were studied, as it seemed to be a compromise between high temperature for the extraction of skatole and androstenone and minimization of lipid oxidation and the creation of degradation products, which could potentially interfere with the detection of boar taint compounds and saturate the sensors in the case of e-noses. Thus, the detection of skatole and androstenone in the headspace will be examined in the next section.

3.1. VOC Profiles Generated through High-Temperature Incubation of Fat

3.1.1. General Understanding of the Generated VOC Profiles

A total of 48 compounds were correctly identified overall in fat samples regardless of their taint (Table 2). The profiles are composed of a large diversity of molecules including, amongst others, alcohols, aldehydes, furanes, and pyridine derivatives. Although some common characteristics are observed between the six different types of profiles obtained, some differences are also observed. These mainly exist between the heating temperatures rather than between the fat types.

Table 2. GC-MS results of VOCs found in the headspace of heated fat. Results are expressed in relative abundance (%, mean ± s.d.). For each molecule, general information such as the match factor of the molecule when compared to the database, its CAS number, as well as the calculated RI and literature RI (NIST17) are given. Finally, relative abundance (%) is given for the six modalities tested: sow back fat heated at 150 °C (S 150 °C) and 180 °C (S 180 °C), untainted back fat heated at 150 °C (UT 150 °C) and 180 °C (UT 180 °C), as well as tainted back fat heated at 150 °C (T 150 °C) and at 180 °C (T 180 °C).

	Match Factor	CAS	Calculated RI	Litt. RI	S 150 °C (n = 6)	S 180 °C (n = 6)	UT 150 °C (n = 7)	UT 180 °C (n = 7)	T 150 °C (n = 7)	T 180 °C (n = 7)
Alcohol										
Pent-1-en-3-ol	87	616-25-1	712	671	0.00 ± 0.00	0.31 ± 0.38	0.2 ± 0.22	0.07 ± 0.18	0.32 ± 0.3	0.08 ± 0.14
Pentan-1-ol	91	71-41-0	769	761	0.14 ± 0.23	0.24 ± 0.46	0.69 ± 0.25	0.35 ± 0.43	0.93 ± 0.5	0.45 ± 0.63
Heptan-1-ol	86	111-70-6	968	960	0.06 ± 0.09	0.37 ± 0.57	0.07 ± 0.08	0.00 ± 0.00	0.04 ± 0.12	0.06 ± 0.15
Oct-1-en-3-ol	93	3391-86-4	978	969	0.25 ± 0.29	0.05 ± 0.11	0.6 ± 0.23	0.00 ± 0.00	0.54 ± 0.48	0.29 ± 0.75
Octan-1-ol	90	111-87-5	1070	1059	0.47 ± 0.29	0.31 ± 0.62	0.43 ± 0.21	0.11 ± 0.3	0.34 ± 0.27	0.00 ± 0.00
Total alcohol					0.91	1.28	1.99	0.53	2.17	0.87
Aldehydes										
3-Methylbutanal	92	590-86-3	701	650	0.00 ± 0.00	0.00 ± 0.00	0.00 ± 0.00	0.09 ± 0.17	0.05 ± 0.09	0.16 ± 0.27
Pentanal	94	110-62-3	719	707	0.71 ± 0.34	0.26 ± 0.64	0.88 ± 0.34	0.57 ± 0.73	0.67 ± 0.57	0.31 ± 0.82
(E)-2-Methylbut-2-enal	86	1115-11-3	759	692	0.11 ± 0.17	0.00 ± 0.00	0.2 ± 0.17	0.00 ± 0.00	0.23 ± 0.21	0.00 ± 0.00
Hexanal	98	66-25-1	795	806	2.22 ± 1.32	1.26 ± 1.89	2.47 ± 1.1	0.49 ± 0.67	3.28 ± 1.37	2.32 ± 1.86
(E)-Hex-2-enal	95	6728-26-3	848	814	0.14 ± 0.23	0.07 ± 0.18	0.26 ± 0.43	0.00 ± 0.00	0.12 ± 0.22	0.13 ± 0.25
Heptanal	96	111-71-7	898	905	0.25 ± 0.29	0.92 ± 2.03	0.4 ± 0.29	0.17 ± 0.22	0.45 ± 0.31	0.22 ± 0.28
(E)-Hept-2-enal	95	18829-55-5	954	913	4.45 ± 1.53	2.2 ± 1.62	3.59 ± 1.06	0.99 ± 1.27	3.76 ± 1.41	0.68 ± 0.89
Benzaldehyde	91	100-52-7	957	982	1.95 ± 1.19	0.45 ± 0.62	0.6 ± 0.84	0.31 ± 0.38	0.81 ± 0.51	0.58 ± 0.76
(E,E)-Hepta-2,4-dienal	93	4313-03-5	995	921	12.14 ± 4.58	3.13 ± 3.98	8.06 ± 2.32	6.48 ± 5.5	7.78 ± 1.6	6.21 ± 4.47
Octanal	91	124-13-0	1001	1005	0.15 ± 0.37	1.07 ± 2.62	0.35 ± 0.4	0.14 ± 0.24	0.41 ± 0.39	0.38 ± 0.53
5-Ethylcyclopent-1-enecarboxaldehyde	83	36431-51-3	1027	1020	0.07 ± 0.16	0.00 ± 0.00	0.12 ± 0.21	0.00 ± 0.00	0.05 ± 0.12	0.12 ± 0.31
Benzeneacetaldehyde	90	122-78-1	1042	1081	0.16 ± 0.1	0.01 ± 0.03	0.13 ± 0.07	0.08 ± 0.11	0.09 ± 0.09	0.07 ± 0.1
(E)-Oct-2-enal	81	2548-87-0	1057	1013	1.52 ± 0.7	0.00 ± 0.00	1.51 ± 0.6	0.37 ± 0.45	1.43 ± 0.36	1.71 ± 3.53
Nonanal	97	124-19-6	1102	1104	2.16 ± 0.86	1.71 ± 1.87	2.16 ± 0.71	1.96 ± 1.51	2.48 ± 0.69	1.85 ± 1.34
(E)-Non-2-enal	96	18829-56-6	1159	1112	0.52 ± 0.47	0.26 ± 0.4	0.56 ± 0.25	0.83 ± 0.79	0.4 ± 0.29	0.00 ± 0.00
(E,E)-Nona-2,4-dienal	93	5910-87-2	1213	1120	0.73 ± 0.31	0.00 ± 0.00	0.64 ± 0.35	0.00 ± 0.00	0.36 ± 0.27	0.00 ± 0.00
(E)-Dec-2-enal	97	3913-81-3	1261	1212	4.26 ± 1.77	3.47 ± 2.87	3.56 ± 1.28	1.9 ± 3.33	2.88 ± 0.96	2.16 ± 4.7
(E,E)-Deca-2,4-dienal	97	25152-84-5	1316	1220	16.34 ± 3.29	4.38 ± 4.27	15.59 ± 4.76	3.14 ± 2.44	10.47 ± 3.51	6.09 ± 4.18
(E)-Undec-2-enal	97	2463-77-6	1363	1311	7.12 ± 2.74	0.17 ± 0.41	6.03 ± 2.58	1.58 ± 1.34	4.4 ± 1.71	3.83 ± 5.07
Hexadecanal	96	629-80-1	1814	1800	0.00 ± 0.00	0.03 ± 0.06	0.26 ± 0.41	0.00 ± 0.00	0.00 ± 0.00	0.00 ± 0.00
Total aldehydes					55.00	19.37	47.37	19.09	40.09	26.81
Alkanes										
n-Heptane	88	142-82-5	719	717	0.00 ± 0.00	0.00 ± 0.00	0.00 ± 0.00	0.14 ± 0.24	0.00 ± 0.00	0.36 ± 0.62
Total alkanes					0.00	0.00	0.00	0.14	0.00	0.36

Table 2. Cont.

	Match Factor	CAS	Calculated RI	Litt. RI	S 150 °C (n = 6)	S 180 °C (n = 6)	UT 150 °C (n = 7)	UT 180 °C (n = 7)	T 150 °C (n = 7)	T 180 °C (n = 7)
Furans										
2-Ethylfuran	87	3208-16-0	722	742	0.12 ± 0.2	0.00 ± 0.00	0.27 ± 0.15	0.03 ± 0.07	0.34 ± 0.35	0.02 ± 0.05
2-Pentylfuran	91	3777-69-3	990	1040	1.69 ± 0.54	0.2 ± 0.36	1.52 ± 0.63	3.79 ± 9.28	1.83 ± 0.63	1.1 ± 1.95
2-[(E)-pent-1-enyl]furan	85	81677-78-3	997	1048	0.00 ± 0.00	0.00 ± 0.00	0.04 ± 0.1	0.00 ± 0.00	0.09 ± 0.2	0.00 ± 0.00
2-Heptylfuran	88	3777-71-7	1190	1239	0.32 ± 0.32	0.00 ± 0.00	0.12 ± 0.23	0.00 ± 0.00	0.00 ± 0.00	0.00 ± 0.00
Total furans					2.13	0.2	1.95	3.82	2.26	1.12
Ketones										
Pentan-3-one	84	96-22-0	681	654	0.00 ± 0.00	0.05 ± 0.08	0.00 ± 0.00	0.00 ± 0.00	0.00 ± 0.00	0.01 ± 0.02
Pentadecan-2-one	95	2345-28-0	1697	1648	4.03 ± 1.12	1.46 ± 1.18	2.65 ± 1.57	1.92 ± 3.24	2.15 ± 1.05	0.7 ± 0.78
Heptadecan-2-one	93	2922-51-2	1900	1847	1.81 ± 0.58	0.45 ± 0.54	0.97 ± 0.71	1.24 ± 1.54	1.67 ± 1.17	0.26 ± 0.7
Total ketones					5.84	1.96	3.62	3.15	3.82	0.97
Acids										
Nonanoic acid	91	112-05-0	1276	1272	0.15 ± 0.37	0.00 ± 0.00	0.27 ± 0.48	0.00 ± 0.00	0.11 ± 0.28	0.16 ± 0.43
Tetradecanoic acid	98	544-63-8	1764	1769	1.44 ± 0.35	0.95 ± 0.91	2.25 ± 0.62	1.23 ± 2.19	2.53 ± 0.56	1.21 ± 1.36
(Z)-Hexadec-11-enoic acid	80	2271-34-3	1903	1886	1.44 ± 1.15	0.00 ± 0.00	2.03 ± 1.75	0.38 ± 1	2.11 ± 2.09	0.00 ± 0.00
Hexadecanoic acid	96	57-10-3	1976	1968	13.65 ± 6.53	15.23 ± 20.87	12.74 ± 4.68	9.82 ± 9.7	11.27 ± 3.41	15.85 ± 19.38
Octadec-9-enoic acid	83	112-79-8	2143	2133	14.46 ± 7.93	31.93 ± 20.34	21.08 ± 8.79	29.51 ± 27.02	29.24 ± 12.83	11.94 ± 13.1
Octadecanoic acid	88	57-11-4	2178	2167	2.59 ± 1.51	11.83 ± 9.51	3.26 ± 1.72	15.54 ± 22.13	3.15 ± 1.77	20.7 ± 30.62
Octadeca-9,12-dienoic acid	84	544-71-8	2283	2183	1.68 ± 0.92	3.2 ± 1.76	2.31 ± 2.34	3.56 ± 5.41	2.56 ± 2.63	3.03 ± 3.47
Total acids					35.41	63.14	43.94	60.03	50.97	52.88
Amides										
Hexadecanamide	92	629-54-9	2192	2021	0.00 ± 0.00	3.38 ± 2.99	0.00 ± 0.00	1.27 ± 1.71	0.00 ± 0.00	3.43 ± 4.05
Octadec-9-enamide	88	301-02-0	2361	2228	0.00 ± 0.00	3.46 ± 1.84	0.00 ± 0.00	3.92 ± 4.68	0.00 ± 0.00	3.05 ± 3.69
Octadecanamide	89	124-26-5	2387	2220	0.09 ± 0.1	0.14 ± 0.34	0.00 ± 0.00	2.05 ± 3.42	0.00 ± 0.00	1.4 ± 1.64
Total amides					0.09	6.97	0.00	7.24	0.00	7.88
Pyridine derivatives										
2-Ethyl-pyridine	89	100-71-0	902	887	0.00 ± 0.00	0.47 ± 0.42	0.00 ± 0.00	0.12 ± 0.21	0.00 ± 0.00	0.17 ± 0.28
2-Pentyl-pyridine	90	2294-76-0	1196	1185	0.43 ± 0.49	0.86 ± 0.72	0.49 ± 0.18	0.29 ± 0.45	0.11 ± 0.19	0.89 ± 0.91
Total pyridines derivates					0.43	1.34	0.49	0.41	0.11	1.06

Table 2. Cont.

	Match Factor	CAS	Calculated RI	Litt. RI	S 150 °C (n = 6)	S 180 °C (n = 6)	UT 150 °C (n = 7)	UT 180 °C (n = 7)	T 150 °C (n = 7)	T 180 °C (n = 7)
Others										
Unknown A			662		0.00 ± 0.00	0.00 ± 0.01	0.00 ± 0.00	0.00 ± 0.00	0.02 ± 0.05	0.01 ± 0.02
Unknown B			1714		0.00 ± 0.00	1 ± 1.56	0.00 ± 0.00	1.14 ± 1.65	0.00 ± 0.00	3.78 ± 2.6
Unknown C			1868		0.2 ± 0.28	0.31 ± 0.6	0.44 ± 0.25	0.06 ± 0.17	0.38 ± 0.27	0.2 ± 0.38
gamma-Palmitolactone	90	730-46-1	2104	1980	0.00 ± 0.00	0.00 ± 0.00	0.2 ± 0.54	3.47 ± 5.32	0.19 ± 0.5	2.85 ± 3.82
delta-hexadecalactone	85	7370-44-7	2133	2000	0.00 ± 0.00	4.16 ± 10.18	0.00 ± 0.00	0.76 ± 2.02	0.00 ± 0.00	1.04 ± 2.76
Squalene	89	111-02-4	2830	2914	0.00 ± 0.00	0.27 ± 0.3	0.00 ± 0.00	0.15 ± 0.19	0.00 ± 0.00	0.18 ± 0.24
Total others					0.2	5.74	0.64	5.59	0.58	8.06
Total					100.00	100.00	100.00	100.00	100.00	100.00

In fact, it can be observed from Table 2 that the major group of compounds identified is not the same at 150 and 180 °C. Aldehydes are the most abundant at 150 °C ranging from 40.09% of the total profile for tainted fats to 55.00% for sow fats compared to much lower percentages of aldehydes at 180 °C, ranging from 19.09% for untainted fat to 26.81% for tainted fats (effect of temperature: p-value < 0.05). Amongst these aldehydes, some are present in much greater quantities compared to others. These include (E)-Dec-2-enal, (E)-Undec-2-enal, (E,E)-Hepta-2,4-dienal, and (E,E)-Deca-2,4-dienal, the latter accounting for up to 16.34% of the total profile in the case of sow fat.

On the other hand, the fatty acids group is the most present at 180 °C, making up 52.88% to 63.14% of the total profile at this temperature. Three fatty acids stand out: octadec-9-enoic acid (up to 31.93% of the total profile), hexadecanoic acid (up to 15.85%), and lastly, octadecanoic acid (up to 20.7%). Finding these three molecules as the most abundant fatty acids is in accordance with what has been found by Zhao et al. (2017) [41], who analyzed VOCs of stewed pork broth by solvent-assisted flavor evaporation (SAFE) combined with GC-MS.

Additionally, observing octadec-9-enoic acid and hexadecanoic acid as two of the three major fatty acids in the VOCs profile corresponds to the actual fatty acids content of back fat. In fact, several studies have analyzed the fatty acid composition of back fat and have found that the most abundant was octadec-9-enoic acid followed by hexadecanoic acid [42,43]. The hydrolysis of triglycerides into free fatty acids (FFAs) and glycerol is controlled by two main lipolytic enzymes: adipose triacylglycerol lipase (ATGL) regulating the hydrolysis of triacylglycerols into diacylglycerols and FFAs and hormone-sensitive lipase (HSL) regulating that of diacylglycerols into monoacylglycerols, FFAs, and glycerols [44]. Therefore, this explains the presence of FFAs in back fat.

Regarding their presence in the headspace, one must remember that such long-chain fatty acids possess low vapor pressures (e.g., octadec-9-enoic acid has a vapor pressure of 5.46×10^{-7} mm Hg at 25 °C [45]); therefore, greater incubation temperatures lead to greater headspace concentrations of these FFAs. With temperatures increasing from 150 to 180 °C, it can be seen from the data that the total acids found in the headspace increase for all three fat types ($p < 0.05$).

Serra et al. (2004) [42] and Rius et al. (2005) [34] who have also analyzed VOCs obtained following incubation of fat observed that aldehydes were the most abundant class of molecules, making up respectively 37.1% and 69.61% of the total VOC profiles. However, lower incubation temperatures (60 and 120 °C) were used in their study, which could explain the smaller volatilization of FFAs and hence the smaller relative abundance of these in their VOC profiles. Seeing that the total aldehydes percentage in the 180 °C profiles is lower is simply due to the fact that more volatiles are being released at 180 °C compared to 150 °C.

As observed in Table 2, the majority of aldehydes present are unsaturated, which is explained by the higher proportions of unsaturated fatty acids than saturated fatty acids in pork back fat [46]. The most abundant aldehydes are (E,E)-deca-2,4-dienal and (E,E)-hepta-2,4-dienal, which are VOCs produced following the oxidation of linoleic acid and linolenic acid, respectively, and which are known to have a fatty and fried smell [24,41,47,48]. Benzaldehyde has also been found to originate from linolenic acid degradation [49].

In smaller proportions are ketones and alcohol. The alcohols detected at these temperatures correspond to those that have been found by Rius et al. (2005) [34] at 120 °C. On the other hand, two of the three ketones (pentadecan-2-one and heptadecan-2-one) observed in our study have not been observed by the latter. However, Zhao et al. (2017) [41] have found pentadecan-2-one as part of the VOCs found in pork broth.

Furans have also been found in the profiles. Furans are well-known to be responsible for the characteristic odor of fried foodstuffs. These molecules are found in a multitude of food products, including meat products [50,51].

3.1.2. Understanding the Differences between the VOC Profiles Generated

Principal component analysis (PCA) was used to better visualize existing differences or groupings between the samples analyzed. Figure 1 represents a PCA scores plot of the first two principal components (PCs) of the VOC profiles dataset. Therefore, in this PCA scores plot, each sample analyzed is represented based on their respective VOC profiles. The first principal component (PC 1) explains 22.4% of the variation in the dataset, while the second principal component (PC 2) explains 7.2% of the variation. In this figure, the samples that are close to each other have similar VOC profiles. Therefore, the clear overlap of sow fat, tainted boar fat, and untainted boar fat respectively at 150 and 180 °C suggests that no net distinction is perceived between the VOC profiles of these three fat types. However, although a slight overlap is observed between the samples at 150 °C and those at 180 °C, a separation exists between the VOC profiles obtained following fat incubation at 150 and 180 °C. This suggests, as expected, that temperature has an impact on the generated VOCs. The molecules majorly responsible for the differences observed between the temperatures are described later in this section (Figure 2).

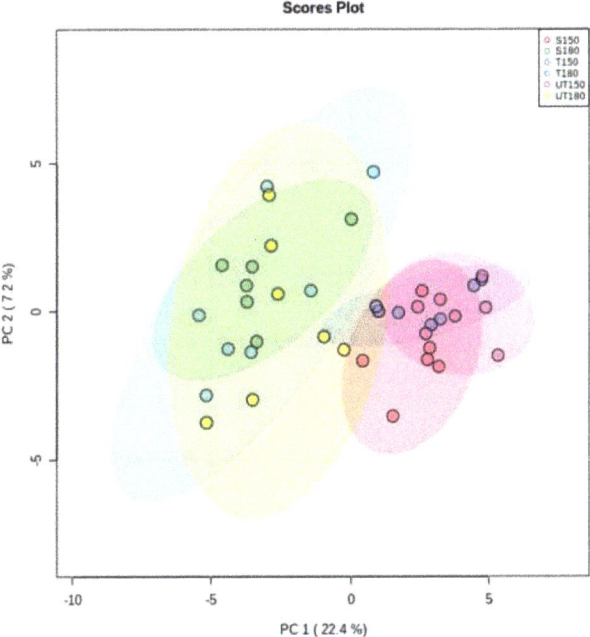

Figure 1. Principal component analysis (PCA) scores plot of component area normalized data of VOC profiles. Red and green dots indicate VOC profiles obtained for sow fat heated at 150 °C ($n = 6$) and 180 °C ($n = 6$), dark blue and light blue represent VOC profiles for tainted boar fat heated at 150 °C ($n = 7$) and 180 °C ($n = 7$); lastly, pink and yellow dots represent untainted fats at 150 °C ($n = 7$) and 180 °C ($n = 7$) respectively.

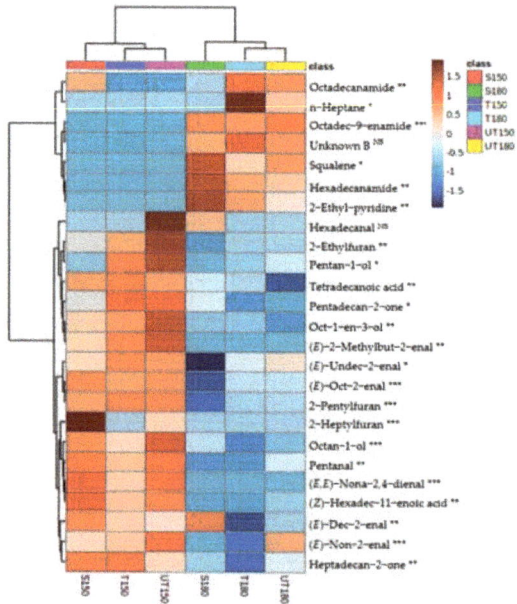

Figure 2. Heatmap generated with normalized data for the top 25 molecules responsible for the differences between the profiles. Each column corresponds to a studied modality. Red and green squares indicate VOC profiles obtained for sow fat heated at 150 and 180 °C, dark blue and light blue represent VOC profiles for tainted boar fat heated at 150 and 180 °C, and lastly, pink and yellow dots represent untainted fats at 150 and 180 °C, respectively. Results of ANOVAs are represented after the molecule name: ^{NS} indicates a p-value > 0.05 while *, **, *** indicate p-values < 0.05, <0.01, and <0.001 respectively.

As a reminder, the general VOC profiles have been established based on an untargeted approach following the detection of molecules in SCAN mode. Hence, skatole and androstenone semi-quantified following SIM mode detection (addressed in the next section) have not been included in the PCA data. Several other molecules have been suggested in the literature as responsible for boar taint. These include indole, 4-phenyl but-3-en-2-one, styrene, 1,4-dichlorobenzene, 2-aminoacetophenone, 5-α-androst-16-en-3-α-ol, and 5-α-androst-16-en-3-β-ol [2,34,52,53]. However, these molecules were not observed in the SCAN data, and no targeted approach (such as the use of the SIM mode) was used to attempt to detect them. Hence, this partially explains the overlapping of tainted and untainted fats. Additionally, although these molecules are not detected here in SCAN mode due to very low analytical concentrations, these still impact sensory evaluation as they may be detected by the human. The concept of odor activity values (OAVs) is very important in such analysis. This one considers the concentration of a compound in the food matrix and its odor threshold. OAV values greater than 1 are considered to be main contributors to the total flavor [54,55]. The OAV in fat of several molecules introduced above have been studied by Gerlach et al. (2018) [56]. For example, they have found that androstenone has an OAV of 25 and skatole has an OAV of 40 in boar fat. On the other hand, (E,E)-deca-2,4-dienal, the most present aldehyde in our study, only had an OAV of 1. This suggests that although this molecule is present in high concentrations in our study (Table 2), it only minorly impacts sensory evaluation compared to boar taint compounds.

The interpretation of the molecules responsible for the difference between the two temperatures is eased through the elaboration of a heatmap (Figure 2). The level of significance of the differences observed can be observed after the molecule name. From the latter, it appears that significant differences exist for 23 of the 25 molecules most responsible for the differences perceived.

As mentioned earlier and as confirmed by this figure, it can be noticed that overall, the differences mainly reside between profiles at the different temperatures. The VOCs' intensities are very different from one temperature to another and imply that assessors performing sensory evaluation at different temperature are not confined to the same working environment. This could lead to different results for the same sample. This stresses the importance of standardizing sensory evaluation protocols, from the training of the assessors to the evaluation per se performed in the slaughterhouse [27,57,58].

Some molecules are present in significantly higher concentrations in the headspace of fat heated at 150 °C compared to 180 °C. For example, this is the case for the aldehydes such as (E)-non-2-enal, (E)-undec-2-enal, and (E,E)-nona-2,4-dienal. As mentioned earlier, these molecules are secondary oxidation products of fatty acids. In meat, these molecules can further react. For example, the molecule (E,E)-deca-2,4-dienal can react with ammonia to produce 2-pentyl-pyridine [46]. Ammonia usually originates from the Strecker degradation of cysteine, which is an amino acid frequently found in the meat [59]. Careful attention was paid when sampling the fat before homogenization; however, the potential traces of muscle (Longissimus dorsi) in the sample cannot be excluded. The acceleration of such a reaction at high temperatures could explain the smaller headspace concentrations of (E,E)-deca-2,4-dienal in samples at 180 °C (Figure 2).

To the exception of octadec-9-enamide in the specific case of sow fat, the fatty amides, hexadecanamide, octadecanamide, and octadec-9-enamide, which are present in the 180 °C profiles, are simply absent from the profiles at 150 °C. Such amides have been obtained in several studies on the pyrolysis of meat products for waste management [60,61], hence demonstrating the implication of high temperatures in their production. These molecules are not simple degradation products of fatty acids and suggest once again the presence of small concentrations of proteins in the fat sample [60].

Another molecule present only in profiles at 180 °C is squalene. This one also has a low vapor pressure (6.3×10^{-6} mmHg at 25 °C), which explains its presence only at the higher temperature. Finding squalene in the three types of fats at this temperature is explained by the fact that squalene affects cholesterol production, which in turn affects the production of the steroid pregnenolone. All steroids and hence both androgens and estrogens in male and female pigs are produced starting from pregnenolone [53].

3.2. Detection of Skatole and Androstenone in the Headspace of Tainted and Untainted Boar Fat

Detection of the ions used for both qualification and semi-quantification of skatole and androstenone in boar fat was possible at both temperatures (Figure 3).

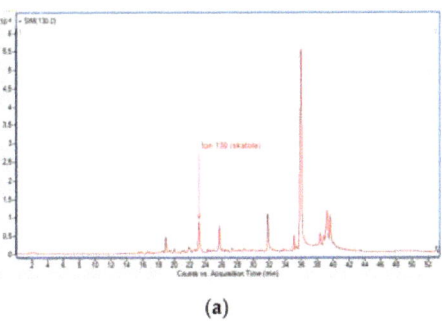
(a)

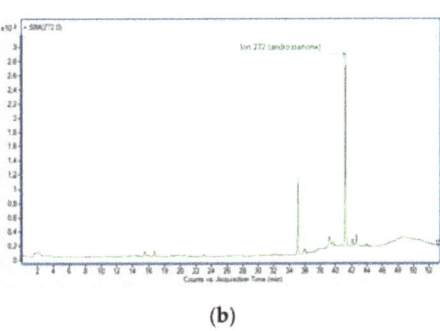
(b)

Figure 3. Detection of ions in selection ion monitoring mode (SIM mode) for samples incubated at 150 °C. Quantitative ions for semi-quantification of (**a**) skatole (*m/z* 130, peak at Rt = 23.096 min) and (**b**) androstenone (*m/z* 272, peak at Rt = 41.233 min).

Additionally, positive correlation coefficients higher than 0.77 are observed between the content (Table S1 for skatole and androstenone back fat content) and headspace concentrations of skatole and androstenone at both 150 and 180 °C (Figure 4).

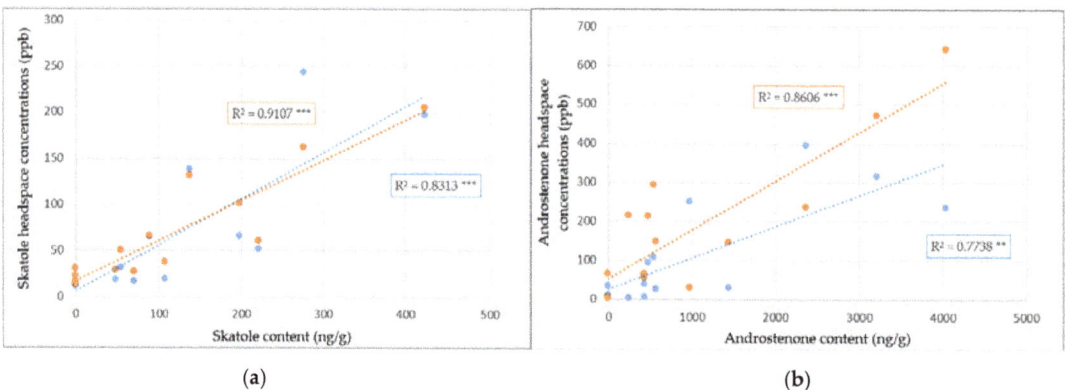

Figure 4. Correlation plots between headspace concentrations (ppb) and content concentrations (ng/g) of skatole (**a**) and androstenone (**b**) respectively at 150 °C (in blue, $n = 7$) and 180 °C (in orange, $n = 7$). For the content, values below the linearity range were set at 0 arbitrarily. Pearson correlation coefficients as well as their significance levels are represented on the graph. **, and *** indicate *p*-values < 0.01, and <0.001, respectively.

It can be observed from the skatole correlation plots that trends between content and emissions are similar at 150 and 180 °C, which therefore suggests that the skatole extraction yield and subsequently the concentrations perceived in the headspace are the same at these temperatures. To increase the headspace concentrations in skatole, several solutions exist. Amongst these solutions, a large increase in temperature, a reduction of the headspace volume, or simply the heating of bigger samples (that hence have a greater absolute quantity in skatole) could be considered. However, as mentioned earlier, in both cases, the VOCs profile will be rich in many other molecules, which might affect the response of the boar taint detection method used (saturation of the assessor's nose in the case of sensory evaluation and sensor drift for sensor-based methods).

Different results appear for androstenone correlation plots. In fact, it can be seen from Figure 4b that more androstenone is emitted with increasing temperature. This can in part be explained by the fact that androstenone has a lower vapor pressure and hence a smaller tendency to volatilize compared to skatole, thus leading to better androstenone extractions at higher temperatures.

Low headspace concentrations (maximum below 250 ppb for skatole and 700 ppb for androstenone) could in part be explained by the strong matrix effects observed with boar fat. Given the lipophilic character of skatole and androstenone, an efficient extraction process is often used prior to analytical determinations of skatole and androstenone. Sample preparation usually begins with a heating or homogenization step followed by an extraction and purification step prior to analysis. Additionally, measurements are often performed based on liquefied fat from which connective tissues have been removed (only 60% of the adipose tissue is constituted of fat per se) [62]. Various methods have been developed in the last decade to quantify skatole and androstenone content based on headspace analysis. As it is the case in our study, these researchers have only incubated fat at high temperatures prior to quantification of the boar taint compound. However, to compensate for matrix effects and subsequent low analyte extraction, internal standards were spiked directly in the liquefied fat [63,64]. This procedure was not performed in our study, as we wanted to determine real headspace concentrations of skatole and androstenone, justifying the injection of the internal standard 2,3-dimethylindole directly in the headspace of the vial. It is important to note that what is perceived by the sensory assessor in the slaughterhouse or by the consumer when eating pork are compounds that are present in gaseous form in the headspace. In fact, VOCs can reach the nasal cavity either through orthonasal olfaction (direct inhalation in front of the nose) or through the throat while chewing (retronasal olfaction) [65].

The positive and significant Pearson correlation coefficients between the headspace concentration and the content for skatole and androstenone indicate that what is found in the headspace at both temperatures is a good representation of boar taint in the fat. Although extraction is similar for skatole at 150 and 180 °C (typical cooking temperatures of pork), the greater headspace concentrations for androstenone at 180 °C imply that higher temperatures allow a better representation of boar taint. One must remember that boar taint is a complex smell composed of a large variety of molecules. Only 33% of boar taint is explained by skatole alone, while 50% of the taint is explained by the combination of skatole and androstenone [66]. Whether it is for sensory evaluation or for sensor-based methods focusing on the detection of skatole and androstenone, using higher temperatures to detect greater amounts of androstenone should allow a better visualization of boar taint as a whole. Lastly, the headspace concentrations are low for sensor-based methods (which often operate in up to the ppm [67]) and hence emphasize the importance of testing even higher temperatures.

4. Conclusions

To the best of our knowledge, this study is the first analyzing VOCs emitted by back fat samples when heated at elevated temperatures. The aim of this study was to perform a general analysis of VOC profiles obtained with fat presenting different boar taint intensities, but being an exploratory study, it did not intend to rigorously compare the impact of different taint combinations on the emitted VOCs. As a reminder, the comprehension of the VOC profiles at these typical cooking temperatures was primordial to understand what composes the exact smell perceived during the sensory evaluation of boar taint at these temperatures and secondly understand whether VOCs sensor-based methods for boar taint detection at these temperatures can be developed.

Great differences were observed between the VOC profiles depending on the incubation temperature. Different VOC profiles might result in differences in classification of the same tainted and untainted fats when heated at different temperatures. Therefore, this stresses the need to develop and use a standardized method for the sensory evaluation of boar taint.

VOCs sensors for skatole and androstenone detection could be developed for incubation temperatures of 150 and 180 °C given that both molecules are found in the headspace. However, the low headspace concentration observed for both these molecules should encourage further research into higher incubation temperatures. Analyses of the general VOCs headspace should always complement research into skatole and androstenone detection as the complexity of the VOCs profile might increase with temperature. The impact of fatty acids and aldehydes (as these are the most abundant in the VOC profiles at both temperatures) should be tested on sensor material to determine the rate at which sensor drift occurs to elaborate more robust drift correction algorithms and finally determine after how many analyses the sensors should be disposed of. Solutions to reduce the development of products from lipid oxidation, such as working in a closed and controlled environment, should be further looked into for sensor development.

Supplementary Materials: The following are available online at https://www.mdpi.com/article/10.3390/foods10061311/s1, Table S1: Quantification of skatole and androstenone in fat determined by HPLC-FD. The mention < LR indicates that the content is below the linearity range (45 to 500 ng/g for skatole and 240 to 5000 ng/g for androstenone).

Author Contributions: Conceptualization, C.B., A.M. and M.-L.F.; methodology, C.B., A.M. and M.-L.F.; software, C.B., A.M. and M.-L.F.; validation, C.B., A.M., M.D., D.L., J.R., A.L. and M.-L.F.; formal analysis, C.B., A.M. and M.-L.F.; investigation, C.B. and A.M.; resources, C.B., A.M. and

M.-L.F.; data curation, C.B., A.M.; writing—original draft preparation, C.B., A.M.; writing—review and editing, M.D., D.L., J.R., A.L. and M.-L.F.; visualization, C.B. and A.M.; supervision, M.-L.F.; project administration, M.-L.F., D.L. and M.D.; funding acquisition, M.-L.F. and M.D. All authors have read and agreed to the published version of the manuscript.

Funding: This research was funded by the European Regional Development Fund (ERDF) and the Walloon Region of Belgium, through the Interreg V France-Wallonie-Vlaanderen program, under the PATHACOV project (No. 1.1.297); and the Micro+ project co-funded by the ERDF and Wallonia, Belgium (No. 675781-642409). This article was written within the framework of the AGROSENSOR project, which is part of the "Pole de compétitivité WAGRALIM," and was financially supported by the "Service public de Wallonie" (SPW).

Data Availability Statement: Data will be available upon request from the corresponding author.

Acknowledgments: We would like to thank Franck Michels, Thomas Bertrand and Danny Trisman for their technical contribution in this work.

Conflicts of Interest: The authors declare no conflict of interest.

References

1. Patterson, R.L.S. 5α-androst-16-ene-3-one:—Compound responsible for taint in boar fat. *J. Sci. Food Agric.* **1968**, *19*, 31–38. [CrossRef]
2. Vold, E. Fleischproduktionseigenschaftenbei Ebernund Kastraten IV: Organoleptische und gaschromatographische Untersuchungen wasserdampffflüchtiger Stoffe des Rückenspeckes von Ebern. *Meld. Norges Landbr.* **1970**, *49*, 1–25.
3. Backus, G.; Higuera, M.; Juul, N.; Nalon, E.; de Briyne, N. Second Progress Report 2015–2017 on the European Declaration on Alternatives to Surgical Castration of Pigs. Available online: https://www.boarsontheway.com/wp-content/uploads/2018/08/Second-progress-report-2015-2017-final-1.pdf (accessed on 10 April 2021).
4. Zamaratskaia, G.; Rasmussen, M.K. Immunocastration of Male Pigs—Situation Today. *Procedia Food Sci.* **2015**, *5*, 324–327. [CrossRef]
5. Haugen, J.E.; van Wagenberg, C.; Backus, G.; Nielsen, B.E.; Borgaard, C.; Bonneau, M.; Panella-Riera, N.; Aluwé, M. A Study on Rapid Methods for Boar Taint Used or Being Developed at Slaughter Plants in the European Union; BoarCheck Project Final Report. Available online: https://ec.europa.eu/food/sites/default/files/animals/docs/aw_prac_farm_pigs_cast-alt_research_boarcheck_20140901.pdf (accessed on 14 April 2021).
6. Burgeon, C.; Debliquy, M.; Lahem, D.; Rodriguez, J.; Ly, A.; Fauconnier, M.L. Past, present, and future trends in boar taint detection. *Trends Food Sci. Technol.* **2021**, *112*, 1–15. [CrossRef]
7. Verplanken, K.; Stead, S.; Jandova, R.; Van Poucke, C.; Claereboudt, J.; Bussche, J.V.; De Saeger, S.; Takats, Z.; Wauters, J.; Vanhaecke, L. Rapid evaporative ionization mass spectrometry for high-throughput screening in food analysis: The case of boar taint. *Talanta* **2017**, *169*, 30–36. [CrossRef] [PubMed]
8. Lund, B.W.; Borggaard, C.; Isak, R.; Birkler, D.; Jensen, K.; Støier, S. High throughput method for quantifying androstenone and skatole in adipose tissue from uncastrated male pigs by laser diode thermal desorption-tandem mass spectrometry. *Food Chem. X* **2021**, *9*. [CrossRef]
9. Auger, S.; Lacoursière, J.; Picard, P. High-Throughput Analysis of Indole, Skatole and Androstenone in Pork Fat by LDTD-MS/MS Positive MRM Transition LDTD-LC-MS/MS System. 2018. Available online: https://phytronix.com/wp-content/uploads/2018/09/AN-1804-High-throughput-analysis-of-Indole-Skatole-and-Androstenone-in-pork-fat-by-LDTD-MSMS.pdf (accessed on 12 April 2021).
10. Borggaard, C.; Birkler, R.; Meinert, L.; Støier, S. At-line rapid instrumental method for measuring the boar taint components androstenone and skatole in pork fat. In Proceedings of the 63rd International Congress of Meat Science and Technology: Nurturing locally, growing globally, Cork, Ireland, 13–18 August 2017; pp. 279–280.
11. Støier, S. A New Instrumental Boar Taint Detection Method. 2019. Available online: https://www.boarsontheway.com/wp-content/uploads/2019/11/Boar-taint-detection-02102019-aangepast.pdf (accessed on 12 April 2021).
12. Liu, X.; Schmidt, H.; Mörlein, D. Feasibility of boar taint classification using a portable Raman device. *Meat Sci.* **2016**, *116*, 133–139. [CrossRef]
13. Sørensen, K.M.; Westley, C.; Goodacre, R.; Engelsen, S.B. Simultaneous quantification of the boar-taint compounds skatole and androstenone by surface-enhanced Raman scattering (SERS) and multivariate data analysis. *Anal. Bioanal. Chem.* **2015**, *407*, 7787–7795. [CrossRef]
14. Hart, J.; Crew, A.; McGuire, N.; Doran, O. *Sensor and Method for Detecting Androstenone or Skatole in Boar Taint (Patent No. EP2966441A1)*; European Patent Application: Munich, Germany, 2016; Available online: https://patents.google.com/patent/EP2966441A1/en (accessed on 10 April 2021).
15. Westmacott, K.L.; Crew, A.P.; Doran, O.; Hart, J.P. Novel, rapid, low-cost screen-printed (bio)sensors for the direct analysis of boar taint compounds androstenone and skatole in porcine adipose tissue: Comparison with a high-resolution gas chromatographic method. *Biosens. Bioelectron.* **2020**, *150*, 111837. [CrossRef]

16. Verplanken, K.; Wauters, J.; Van Durme, J.; Claus, D.; Vercammen, J.; De Saeger, S.; Vanhaecke, L. Rapid method for the simultaneous detection of boar taint compounds by means of solid phase microextraction coupled to gas chromatography/mass spectrometry. *J. Chromatogr. A* **2016**, *1462*, 124–133. [CrossRef]
17. Sørensen, K.M.; Engelsen, S.B. Measurement of boar taint in porcine fat using a high-throughput gas chromatography-mass spectrometry protocol. *J. Agric. Food Chem.* **2014**, *62*, 9420–9427. [CrossRef] [PubMed]
18. Berdague, J.; Talou, T. Examples of semiconductor gas sensors applied to meat products. *Sci. Aliment.* **1993**, *13*, 141–148.
19. Bourrounet, B.; Talou, T.; Gaset, A. Application of a multi-gas-sensor device in the meat industry for boar-taint detection. *Sens. Actuators B Chem.* **1995**, *27*, 250–254. [CrossRef]
20. Annor-Frempong, I.E.; Nute, G.R.; Wood, J.D.; Whittington, F.W.; West, A. The measurement of the responses to different odour intensities of "boar taint" using a sensory panel and an electronic nose. *Meat Sci.* **1998**, *50*, 139–151. [CrossRef]
21. Di Natale, C.; Pennazza, G.; Macagnano, A.; Martinelli, E.; Paolesse, R.; D'Amico, A. Thickness shear mode resonator sensors for the detection of androstenone in pork fat. *Sens. Actuators B Chem.* **2003**, *91*, 169–174. [CrossRef]
22. Vestergaard, J.S.; Haugen, J.E.; Byrne, D.V. Application of an electronic nose for measurements of boar taint in entire male pigs. *Meat Sci.* **2006**, *74*, 564–577. [CrossRef]
23. Ladikos, D.; Lougovois, V. Lipid Oxidation in Muscle Foods: A Review. *Food Chem.* **1990**, *35*, 295–314. [CrossRef]
24. Domínguez, R.; Pateiro, M.; Gagaoua, M.; Barba, F.J.; Zhang, W.; Lorenzo, J.M. A comprehensive review on lipid oxidation in meat and meat products. *Antioxidants* **2019**, *8*, 429. [CrossRef]
25. Vergara, A.; Vembu, S.; Ayhan, T.; Ryan, M.A.; Homer, M.L.; Huerta, R. Chemical gas sensor drift compensation using classifier ensembles. *Sens. Actuators B Chem.* **2012**, *166–167*, 320–329. [CrossRef]
26. Mortensen, A.B.; Sørensen, S.E. Relationship between boar taint and skatole determined with a new analysis method. In Proceedings of the 30th International Congress of Meat Science and Technology, Bristol, England, 9–14 September 1984; pp. 394–396.
27. Trautmann, J.; Meier-Dinkel, L.; Gertheiss, J.; Mörlein, D. Boar taint detection: A comparison of three sensory protocols. *Meat Sci.* **2016**, *111*, 92–100. [CrossRef] [PubMed]
28. Whittington, F.M.; Zammerini, D.; Nute, G.R.; Baker, A.; Hughes, S.I.; Wood, J.D. Comparison of heating methods and the use of different tissues for sensory assessment of abnormal odours (boar taint) in pig meat. *Meat Sci.* **2011**, *88*, 249–255. [CrossRef]
29. Byrne, D.V.; Thamsborg, S.M.; Hansen, L.L. A sensory description of boar taint and the effects of crude and dried chicory roots (Cichorium intybus L.) and inulin feeding in male and female pork. *Meat Sci.* **2008**, *79*, 252–269. [CrossRef] [PubMed]
30. Hansen, L.L.; Stolzenbach, S.; Jensen, J.A.; Henckel, P.; Hansen-Møller, J.; Syriopoulos, K.; Byrne, D.V. Effect of feeding fermentable fibre-rich feedstuffs on meat quality with emphasis on chemical and sensory boar taint in entire male and female pigs. *Meat Sci.* **2008**, *80*, 1165–1173. [CrossRef]
31. Parunovic, N.; Petrovic, M.; Matekalo-Sverak, V.; Parunovic, J.; Radovic, C. Relationship between carcass weight, skatole level and sensory assessment in fat of different boars. *Czech. J. Food Sci.* **2010**, *28*, 520–530. [CrossRef]
32. Bekaert, K.M.; Aluwé, M.; Vanhaecke, L.; Heres, L.; Duchateau, L.; Vandendriessche, F.; Tuyttens, F.A.M. Evaluation of different heating methods for the detection of boar taint by means of the human nose. *Meat Sci.* **2013**, *94*, 125–132. [CrossRef]
33. Font-i-Furnols, M. Consumer studies on sensory acceptability of boar taint: A review. *Meat Sci.* **2012**, *92*, 319–329. [CrossRef]
34. Rius, M.A.; Hortós, M.; García-Regueiro, J.A. Influence of volatile compounds on the development of off-flavours in pig back fat samples classified with boar taint by a test panel. *Meat Sci.* **2005**, *71*, 595–602. [CrossRef]
35. Bonneau, M. Use of entire males for pig meat in the European union. *Meat Sci.* **1998**, *49*, 257–272. [CrossRef]
36. Hansen-Møller, J. Rapid high-performance liquid chromatographic method for simultaneous determination of androstenone, skatole and indole in back fat from pigs. *J. Chromatogr. B Biomed. Sci. Appl.* **1994**, *661*, 219–230. [CrossRef]
37. Nea, F.; Tanoh, E.A.; Wognin, E.L.; Kenne Kemene, T.; Genva, M.; Saive, M.; Tonzibo, Z.F.; Fauconnier, M.L. A new chemotype of Lantana rhodesiensis Moldenke essential oil from Côte d'Ivoire: Chemical composition and biological activities. *Ind. Crops Prod.* **2019**, *141*, 111766. [CrossRef]
38. Tanoh, E.A.; Boué, G.B.; Nea, F.; Genva, M.; Wognin, E.L.; Ledoux, A.; Martin, H.; Tonzibo, Z.F.; Frederich, M.; Fauconnier, M.L. Seasonal effect on the chemical composition, insecticidal properties and other biological activities of zanthoxylum leprieurii guill. & perr. essential oils. *Foods* **2020**, *9*, 550. [CrossRef]
39. Werrie, P.; Burgeon, C.; Jean, G.; Goff, L.; Hance, T. Biopesticide Trunk Injection Into Apple Trees: A Proof of Concept for the Systemic Movement of Mint and Cinnamon Essential Oils. *Front. Plant. Sci.* **2021**, *12*, 1–13. [CrossRef] [PubMed]
40. Pang, Z.; Chong, J.; Li, S.; Xia, J. MetaboAnalystR 3.0: Toward an Optimized Workflow for Global Metabolomics. *Metabolites.* **2020**, *10*, 186. [CrossRef] [PubMed]
41. Zhao, J.; Wang, M.; Xie, J.; Zhao, M.; Hou, L.; Liang, J.; Wang, S.; Cheng, J. Volatile flavor constituents in the pork broth of black-pig. *Food Chem.* **2017**, *226*, 51–60. [CrossRef] [PubMed]
42. Serra, A.; Buccioni, A.; Rodriguez-Estrada, M.T.; Conte, G.; Cappucci, A.; Mele, M. Fatty acid composition, oxidation status and volatile organic compounds in "Colonnata" lard from Large White or Cinta Senese pigs as affected by curing time. *Meat Sci.* **2014**, *97*, 504–512. [CrossRef]
43. Mörlein, D.; Tholen, E. Fatty acid composition of subcutaneous adipose tissue from entire male pigs with extremely divergent levels of boar taint compounds—An exploratory study. *Meat Sci.* **2014**, *99*, 1–7. [CrossRef]

44. Li, Y.; Zheng, X.; Liu, B.; Yang, G. Regulation of ATGL expression mediated by leptin in vitro in porcine adipocyte lipolysis. *Mol. Cell. Biochem.* **2010**, 121–128. [CrossRef]
45. Oleic Acid I C18H34O2—PubChem. Available online: https://pubchem.ncbi.nlm.nih.gov/compound/Oleic-acid#section=Vapor-Pressure (accessed on 14 April 2021).
46. Mottram, D.S. Flavour formation in meat and meat products: A review. *Food Chem.* **1998**, *62*, 415–424. [CrossRef]
47. Schaich, K.M. Challenges in Elucidating Lipid Oxidation Mechanisms: When, Where, and How Do Products Arise? In *Lipid Oxidation: Challenges in Food Systems*; AOCS Press: Champaign, IL, USA, 2013; pp. 1–52. ISBN 9780988856516.
48. Zang, M.; Wang, L.; Zhang, Z.; Zhang, K.; Li, D.; Li, X.; Wang, S.; Chen, H. Changes in flavour compound profiles of precooked pork after reheating (warmed-over flavour) using gas chromatography–olfactometry–mass spectrometry with chromatographic feature extraction. *Int. J. Food Sci. Technol.* **2020**, *55*, 978–987. [CrossRef]
49. Elmore, J.S.; Mottram, D.S. The role of lipid in the flavour of cooked beef. *Dev. Food Sci.* **2006**, *43*, 375–378. [CrossRef]
50. Yu, A.N.; Sun, B.G.; Tian, D.T.; Qu, W.Y. Analysis of volatile compounds in traditional smoke-cured bacon(CSCB) with different fiber coatings using SPME. *Food Chem.* **2008**, *110*, 233–238. [CrossRef]
51. Roldán, M.; Ruiz, J.; del Pulgar, J.S.; Pérez-Palacios, T.; Antequera, T. Volatile compound profile of sous-vide cooked lamb loins at different temperature-time combinations. *Meat Sci.* **2015**, *100*, 52–57. [CrossRef] [PubMed]
52. Fischer, J.; Gerlach, C.; Meier-Dinkel, L.; Elsinghorst, P.W.; Boeker, P.; Schmarr, H.G.; Wüst, M. 2-Aminoacetophenone—A hepatic skatole metabolite as a potential contributor to boar taint. *Food Res. Int.* **2014**, *62*, 35–42. [CrossRef]
53. Brooks, R.I.; Pearson, A.M. Steroid hormone pathways in the pig, with special emphasis on boar odor: A review. *J. Anim. Sci.* **1986**, *62*, 632–645. [CrossRef] [PubMed]
54. Han, D.; Zhang, C.H.; Fauconnier, M.L. Effect of seasoning addition on volatile composition and sensory properties of stewed pork. *Foods* **2021**, *10*, 83. [CrossRef]
55. Han, D.; Zhang, C.H.; Fauconnier, M.L.; Jia, W.; Wang, J.F.; Hu, F.F.; Xie, D.W. Characterization and comparison of flavor compounds in stewed pork with different processing methods. *LWT* **2021**, *144*, 111229. [CrossRef]
56. Gerlach, C.; Leppert, J.; Santiuste, A.C.; Pfeiffer, A.; Boeker, P.; Wüst, M. Comparative Aroma Extract Dilution Analysis (cAEDA) of Fat from Tainted Boars, Castrated Male Pigs, and Female Pigs. *J. Agric. Food Chem.* **2018**, *66*, 2403–2409. [CrossRef]
57. Heyrman, E.; Janssens, S.; Buys, N.; Vanhaecke, L.; Millet, S.; Tuyttens, F.A.M.; Wauters, J.; Aluwé, M. Developing and understanding olfactory evaluation of boar taint. *Animals* **2020**, *10*, 1684. [CrossRef]
58. Font-i-Furnols, M.; Martín-Bernal, R.; Aluwé, M.; Bonneau, M.; Haugen, J.E.; Mörlein, D.; Mörlein, J.; Panella-Riera, N.; Škrlep, M. Feasibility of on/at line methods to determine boar taint and boar taint compounds: An overview. *Animals* **2020**, *10*, 1886. [CrossRef]
59. Farmer, L.J.; Mottram, D.S. Interaction of lipid in the maillard reaction between cysteine and ribose: The effect of a triglyceride and three phospholipids on the volatile products. *J. Sci. Food Agric.* **1990**, *53*, 505–525. [CrossRef]
60. Leon, M.; Garcia, A.N.; Marcilla, A.; Martinez-Castellanos, I.; Navarro, R.; Catala, L. Thermochemical conversion of animal by-products and rendering products. *Waste Manag.* **2018**, *73*, 447–463. [CrossRef] [PubMed]
61. Berruti, F.M.; Ferrante, L.; Briens, C.L.; Berruti, F. Pyrolysis of cohesive meat and bone meal in a bubbling fluidized bed with an intermittent solid slug feeder. *J. Anal. Appl. Pyrolysis* **2012**, *94*, 153–162. [CrossRef]
62. Haugen, J.E.; Brunius, C.; Zamaratskaia, G. Review of analytical methods to measure boar taint compounds in porcine adipose tissue: The need for harmonised methods. *Meat Sci.* **2012**, *90*, 9–19. [CrossRef] [PubMed]
63. Fischer, J.; Elsinghorst, P.W.; Bücking, M.; Tholen, E.; Petersen, B.; Wüst, M. Development of a candidate reference method for the simultaneous quantitation of the boar taint compounds androstenone, 3α-androstenol, 3β-androstenol, skatole, and indole in pig fat by means of stable isotope dilution analysis-headspace solid-phase micro. *Anal. Chem.* **2011**, *83*, 6785–6791. [CrossRef]
64. Fischer, J.; Haas, T.; Leppert, J.; Schulze Lammers, P.; Horner, G.; Wüst, M.; Boeker, P. Fast and solvent-free quantitation of boar taint odorants in pig fat by stable isotope dilution analysis-dynamic headspace-thermal desorption-gas chromatography/time-of-flight mass spectrometry. *Food Chem.* **2014**, *158*, 345–350. [CrossRef] [PubMed]
65. Genva, M.; Kemene, T.K.; Deleu, M.; Lins, L.; Fauconnier, M.L. Is it possible to predict the odor of a molecule on the basis of its structure? *Int. J. Mol. Sci.* **2019**, *20*, 3018. [CrossRef]
66. Hansson, K.-E.; Lundström, K.; Fjelkner-Modif, S.; Persson, J. The importance of androstenone and skatole for boar taint. *Swedish J. Agric. Res.* **1980**, *10*, 167–173.
67. Spinelle, L.; Gerboles, M.; Kok, G.; Persijn, S.; Sauerwald, T. Review of portable and low-cost sensors for the ambient air monitoring of benzene and other volatile organic compounds. *Sensors* **2017**, *17*, 1520. [CrossRef] [PubMed]

Article

Quality Characteristics of Beef Patties Prepared with Octenyl-Succinylated (Osan) Starch

Mohamed F. Eshag Osman, Abdellatif A. Mohamed *, Mohammed S. Alamri, Isam Ali Mohamed Ahmed, Shahzad Hussain, Mohamed I. Ibraheem and Akram A. Qasem

Department of Food Science and Nutrition, King Saud University, Riyadh 1145, Saudi Arabia; moh.fareed77@yahoo.com (M.F.E.O.); msalamri@ksu.edu.sa (M.S.A.); iali@ksu.edu.sa (I.A.M.A.); shhussain@ksu.edu.sa (S.H.); mfadol@ksu.edu.sa (M.I.I.); aqasem@ksu.edu.sa (A.A.Q.)
* Correspondence: abdmohamed@ksu.edu.sa

Abstract: Octenyl-succinylated corn starch (Osan) was used to improve the physicochemical properties of ground beef patties. The study involved incorporation of 5 and 15% Osan and storage for 30 or 60 days at $-20\,°C$. The tested parameters included cooking loss, microstructure image, firmness, color, and sensory evaluation of the prepared patties. Along with Osan, native corn starch was used as control and considered the patties with added animal fat. The data showed that Osan reduced the cooking loss and dimensional shrinkage significantly ($p < 0.05$), whereas the moisture retention, firmness and color of beef patties were improved. The sensory evaluation indicated enhanced tenderness and juiciness without significant alteration of flavor, color, and overall acceptability of the cooked patties. Microstructure images of cooked patties indicated uniform/cohesive structures with small pore size of patties shaped with Osan. Obviously, good storability of the uncooked patties was reflected on the physiochemical, textural, color, and sensory evaluation of the cooked patties, which points to the benefit of using Osan in frozen patties and signifies possible use in the meat industry. The overall sensory acceptability scores were given to cooked patties containing Osan starch as well as the native starch, whereas 15% animal fat was favored too.

Keywords: beef patties; corn starch; Osan; tenderness; cooking loss

1. Introduction

Meat patties are considered the most popular ready to eat food, due to its desirable sensory and mouth feel. This qualifies its consumption to be considered as a good part of human diet in the past few decades, in addition to its nutritional value which includes essential amino acids, vitamins, and minerals [1]. In processed meat, animal fat is a principal ingredient, found in lumps and plays a central functional and sensory role such as a binding and flavoring agent [2]. However, fat content is a critical obstacle that needs to be addressed especially at high levels. Specifically, fat has been associated with chronic deceases, such as obesity and high cholesterol which leads to hypertension and cardiovascular diseases [3]. Therefore, reducing fat turned into a current trend for the meat-processing industry to meet the consumer demands [4]. Heretofore, countless efforts have been devoted to address this concern in terms of preparing healthy products without reducing the characteristics of the final full fat product such as the physical appearance and sensory qualities. Carbohydrate-based fat replacers, such as inulin, gums, cellulose derivatives, and starches are wildly considered as fat replacer and/or fat substitute applied in low fat meat products [5,6], because of its abundancy, superior functionality, and cost-effectiveness [7,8]. Among these carbohydrates, starch exhibited extremely good functional properties improving viscosity, solubility, and water-holding capacity, as well as, adaptability [9].

Because of its poor water solubility and high retrogradation, native starches utilization is limited. This presented the need for physical, chemical, or enzymatic modifications of native starches so as to expand its utilization [10]. Starch modification can improve its functionality such as viscosity, become more tolerant to various processing conditions such as extreme pH, high temperatures, and shear. For instance, chemically modified starches exhibit significant change in its functional properties compared to physical or enzyme modifications. One of the common chemically modified starches is octenyl-succinylation (Osan) which accrues by esterification of some OH groups by octenyl molecules [11]. To overcome the hydrophilic nature of the abundant hydroxyl groups, various starch modifications by introduction of hydrophobic moiety have been reported. Therefore, amphiphilic modification of starch is one of the methods widely used to improve their hydrophobicity, because amphiphilic starches have a wide range of applications, mainly in emulsification and encapsulation. The amphiphilicity of octenyl succinic anhydride (OSA)-modified starch is improved due to the introduction of dual functional hydrophilic and hydrophobic groups [12]. The obtained Osan starch was utilized in various application ranging from pharmaceutical to food products, such as, puddings, sauces, and baby foods [13]. The use of sodium octenyl succinate starch in a methacrylate/polysaccharides blends, introduced good flowability, surface-active, smoother particle surface, and low-viscosity to spray dried emulsion of the blends [14]. In addition, Osan starch has been used as a fat substitute because it enhances the firmness and high palatability of some meat products [15]. In contrast, octenyl-modified waxy maize starch was used successfully in low fat mayonnaise at substitution level up to 75%, where the product exhibited great sensory quality such as texture and aroma [13]. In baked products, Osan starch improved dough machinability and handling as well as the loaf volume [16].

The objective of this work is to compare the quality characteristics of beef patties prepared with starch to those with animal fat. Therefore, ocetinyle succenylated and native corn starch as fat replacement in beef patties was utilized. This work includes the effect of storability at −20 °C on the quality characteristics of the patties (microstructure, cooking properties, moisture retention, texture, color, and sensory characteristics of the produced patties).

2. Materials and Methods

2.1. Materials

Lean beef meat and animal fat (Beef) were purchased from a central meat market (Al-Taamir Market, Riyadh, Saudi Arabia). Ground lean meats were selected from animals of the same age, breed, and feeding protocol (aldanon farm). While, fresh fat was obtained from Riyadh slaughterhouse (Riyadh, Saudi Arabia). Corn starch was purchased from Middle East Food Solutions Company (Riyadh, Saudi Arabia). Black paper, white paper, onion powder, garlic powder, were purchased from Panda Retail Company (Riyadh, Saudi Arabia). Analytical grade reagents, HCL, 2-Octen-1-yl succinic anhydride, NaOH were purchased from Sigma-Aldrich Chemical Co. (St. Louis, MO, USA).

2.2. Methods

2.2.1. Octenyl-Succinylation of Corn Starch (Osan)

Osan corn starch was synthesized via esterification according to the method of [17] with slight modification. Starch, (500 g) was suspended in 1125 mL of distill water. The pH of the slurry was adjusted to 8.5–9 using 1% NAOH, followed by the addition of Osan (4% based on starch dry weight). The 2-Octen-1-yl succinic anhydride was added slowly with continuous agitation at 35 °C, while maintaining the pH around 8.5–9.0 the reaction was allowed to continue for one hour and the final pH was adjusted to 6.5 using 1.0 N HCL. The obtained mixture was centrifuged at (4000× g) for 10 min, washed twice with distill water and once with acetone. The product was dried at room temperature for two days, ground, sieved through 250 μm sieve and stored for further use. This product is considered as food grade according to the followed preparation method.

2.2.2. Fourier Transform Infrared Spectroscopy (FT-IR)

The stretching vibrational mode of the functional group (Osan) on the modified starch was detected by FTIR spectrophotometer (Bruker, ALPHA, Hanau Germany). One gram of dry sample was placed on the FT-IR cell and scanned between 4000 and 500 cm^{-1}.

2.2.3. Preparation and Processing of Beef Patties

Beef patties were prepared and processed as described by Alejandre et al. [18] with slight modification. Round beef cut was sliced to small pieces and the intramuscular fat was removed. The obtained lean meat was ground to passed through a 5-mm plate meat grinder EMG-1600R ELEKTA ltd, Elekta (Hong Kong, China). Simultaneously, the beef caul fat was melted for 5 min in the microwave oven (sharp, 1200 W output, (Osaka, Japan). Six formulations with the following common ingredients g/100 g addition based on ground lean meat weight (100 g), where 1.0 g of the following ingredients were added salt, hot pepper, white pepper, dry powdered garlic, dry onion powder, and 0.35 g of vinegar, in addition to 20 g of ice water. The experimental design included two treatments and three subsamples; two levels of caul fat, two levels of native corn starch, and two levels of Osan corn starch, hence, the one level included 5% and the second was 15%. Samples with caul fat are considered the control and the melted caul fat was added to the spiced lean ground beef drop-wise with constant hand mixing, but in the starch-containing formulation the caul fat was totally replaced by modified or native starch which was added in small amounts while mixing using a Stephan UM 12 mixer (Stephan U. Sohner GmbH and Co., Gackenbach, Germany). Patties (100 g) were prepared using a patty-making machine (Expro. Co., Shanghai, China). The compressed patties, 100 mm diameter and 10 mm thick, were packaged in vacuumed plastic bags and stored at -20 °C for 0, 30, and 60 days. Before analysis, frozen un-cooked patties were placed at room temperature for 1.5 h. The raw patties were cooked in two steps; the first step was to prepare precooked patties by steaming for 20 min to stabilize the diameter, whereas the second step was the final cooking using the electric hot plate (Stilfer model, 0040, Genova, Italy) for a total of 10 min with 5 min on each side at 180 ± 1 °C. The internal temperature of the patty was 75 °C measured at the geometrical center of the patties using digital thermocouple probe (Ecoscan Temp JKT, Eutech instruments, Pte Ltd., Keppel Bay, Harbour Front, Singapore) The sensory evaluation of the cooked products was carried out directly after cooking.

2.2.4. Scanning Electron Microscopy (SEM)

Microstructures of the obtained cooked patties was examined using JEOL-6360A SEM (Jeol Ltd., Tokyo, Japan). Samples were cut into small pieces $5 \times 5 \times 1$ mm, mounted on the pin stubs using copper tape before coating with gold using an automated sputter coater JFC-1600 Auto Fine Coater (Jeol Ltd., Tokyo, Japan) for 5 min at 2.5 kV operation energy. Subsequently, four fields of each sample were spotted and the selected images were captured at magnifications ranging from $100\times$ to $1000\times$.

2.2.5. Measurement of Cooking Parameters
Cooking Loss

The cooking loss of patties was determined by weighing before and after cooking as recommended by Hollenbeck et al. [19] using the following equation:

$$\text{Cooking loss} = ((\text{un} - \text{cooked patties weight}) - (\text{cooked patties weight}))/(\text{un} - \text{cooked patties weight}) * 100$$

Moisture Content

The moisture contents of un-cooked and cooked patties were determined based on the AOAC Method 950.46 [20].

Moisture Retention

The moisture retention, was determined as the amount of moisture retained in the cooked product per 100 g of raw sample. This value was calculated according to the following equation described by [21,22], where the weight of the patties was recorded before and after cooking, and the cooking yield was calculated by dividing the weight of cooked patties by the weight of uncooked patties and expressed in as reported by [23].

$$\text{Moisture retention} = \text{cooking yield} \times \frac{\text{moisture in coocked patties}}{\text{moisture in un-cooked patties}}$$

$$\text{Cooking yield} = \text{weight before cooking} - \text{cooked weight}$$

Patties Diameter, Thickness, and Shrinkage

Change in beef patties' diameter was determined before and after cooking using Digital Electric Caliper (Pen Tools Co., Maplewood, NJ, USA) by employing the following equations.

$$\text{Diameter} = \frac{(\text{un-cooked patties diameter}) - (\text{cooked patties diameter})}{\text{un-cooked patties diameter}} \times 100$$

$$\text{Thickness} = \frac{(\text{un-cooked patties thickness}) - (\text{cooked patties thickness})}{\text{un-cooked patties thickness}} \times 100$$

$$\text{Dimensional shrinkage} = \frac{(\text{Raw thickness} - \text{cooked thickness}) + (\text{Raw diameter} - \text{cooked diameter})}{\text{Raw thickness} - \text{raw diameter}} \times 100$$

Firmness

The firmness of cooked beef patties was determined using a texture analyzer (TA XT Express, Micro Systems Ltd., Surrey, UK). Samples (60 mm diameter and 10 mm thickness) were pressed using aluminum cylinder probe (SMS P/20 mm diameter, TA XT Plus Micro Systems Ltd., Surrey, UK) operated at 1 mm/s. Samples were compressed to 8 mm distance with 10% strain, where the needed force is expressed in Newton (N). The shear force corresponds to maximum peak force, expressed in Newton (N). The test was performed at room temperature (25 ± 1 °C).

pH

The pH was measured using a portable pH-meter (Model pH 211, Hanna Instruments, Woonsocket, RI, USA) by injecting the probe in 25 g of meat patty and held for 10 s to obtain the pH value.

Surface Color Measurement

The surface color characteristics of un-cooked and cooked patties were determined after the specified storage time (0, 30, or 60) days. The measurements included, lightness (L*), redness (a*), and yellowness (b*), assessed using a portable colorimeter (Konica Minolta, CR-400-Japan; Measuring aperture: 8 mm; Illuminant: CIE D65; Observer angle: CIE 2° Standard Observer). Five color measurements were done, where each patty was separated into four quarters one measurement on the surface of each quarter was taken and the fifth was done in the middle.

2.2.6. Sensory Evaluation of Meat Patties

The sensory test was performed using a 9 points hedonic test (Affective Tests), which includes scale from 1 (dislike extremely) to 9 (like extremely) in a single session. This test is useful for evaluating the acceptance of new products. The sensory evaluation team included trained students and King Saud University staff average age between 22 and

60 years old. The panelists were trained to be able to evaluate the sensory properties of patties including overall product acceptability according to method of [24]. After training, 13 panelists were selected based on their ability and sensitivity to point out differences between the parameters. Cooked patties were tested on a 9-points scale method. The test was conducted in a designated sensory evaluation laboratory with appropriate setting such as partitioned cabinets and individual lightning at 20 ± 2.0 °C. Six treatments with two levels of fat content (control), native or modified starch were evaluated. Patties were cooked as described above and cut into triangles (25×20 mm) and served warm to the participants. Water and mint were also provided to neutralize the flavor between samples [25]. The expert panelists were asked to evaluate the color, flavor, tenderness, juiciness, and overall acceptability.

2.2.7. Statistical Analysis

The statistical analysis was carried out using the Tukey HSD test (Statistic 10 Data analysis software, Inc., Chicago, IL, USA) at ($p \leq 0.05$). The significance level of the analysis of variance (ANOVA) was applied to observe the differences. All measurements were done in triplicate.

3. Results and Discussion

3.1. Fourier Transform Infrared Spectroscopy (FT-IR)

FT-IR profile is shown in Figure 1. In general, the highly intense peaks noticed around 3430 cm^{-1} were ascribed to (O-H) characteristics stretching vibration of amylose or amylopectin, while peaks around 2930 and 1645 cm^{-1} are attributed to C–H stretching and to the tightly bound water present in the starch, respectively [26]. In addition, the peak at 1020 cm^{-1} was originated from the C–O stretching vibration of glucose monomer (Garcia and Grossmann 2014). Two new peaks emerged after modification. Evidently, the region between 1720 and 1570 cm^{-1} is considered a finger print for the main functional groups of the octenyl-succinylated (Osan) corn starch [27]. The new peak at 1571 cm^{-1} emerged after OSA modification was ascribed to the asymmetric stretching vibration of carboxylate RCOO–, whereas the other new peak at 1725 cm^{-1} was observed, which can be attributed to the characteristic C=O stretching vibration of an ester carbonyl group [28].

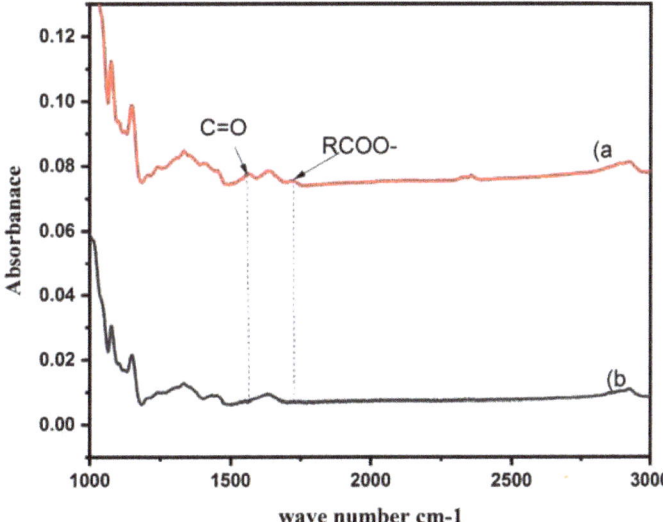

Figure 1. FT-IR spectra of (**a**) Osan-corn starch and (**b**) native corn starch.

3.2. Scanning Electron Microscopy (SEM)

The gel network of patties with 5% addition of either fat or starch, was loose and irregular, because large holes emerged within the structure compared to the 15% (Figure 2). Nonetheless, the gel network structure of the 15% Osan starch was more compact and dense. Furthermore, smaller holes were observed in the surface of the patties with native starch and the holes were more obvious with the increase in the amount of native starch in the samples. Moreno et al. [29] reported better surface structure using a muscle homogenizer than samples with added sodium alginate as a cold gelation technology. Tseng et al. [30] reported dense SEM images of meat balls treated with TGase enzyme compared to untreated samples which indicates rise in the formation of intermolecular ε (γ-glutamyl)-lisil cross-links due to the action of the enzyme. In this study, starch generated a dense network by absorbing the excess water released by the meat during cooking which leads to swelling and closing of the gaps created during cooking. Therefore, smoother denser surface was formed in the presence of starch and more so Osan starch compared to the control. Other researchers used plant material rich in hydrofoils reported improved microstructure of meat and homogenous network [31].

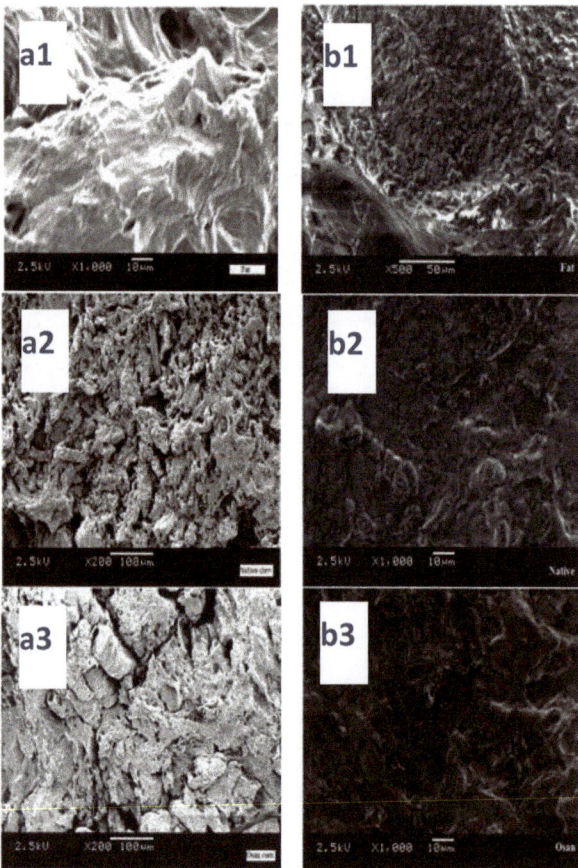

Figure 2. Scanning electron microscope (SEM) images of cooked patties containing animal fat, native corn starch and Osan corn starch (**a1**) control, (**a2**) native starch, and (**a3**) Osam starch 15% addition; (**b1**) control, (**b2**) native starch, and (**b3**) Osan 5% addition.

3.3. Physical Properties of Beef Patties

3.3.1. Cooking Properties

The cooking properties of the patties prepared with Osan-corn starch, native corn, and control are presented in Table 1. The addition of Osan-corn starch at 5 or 15%, significantly ($p \leq 0.05$) altered the cooking loss, moisture retention, thickness, diameter, and dimensional shrinkage of the patties. These parameters were compared to the raw patties. The data presented here are in agreement with the literature reports on the effectivity of corn starch to retain moisture of the cooked bologna [10,32,33]. The firmness of patties prepared with native starch was significantly higher than Osan starch which can be attributed to the amylose retrogradation which is less in Osan starch, thereby, patties with more starch exhibited firmer texture (Table 1) [34]. The firmness of the control was much higher at 5% fat content compared to 15% indicating softer texture due to higher fat content but, it remains significantly higher than those with either type of starch, especially Osan starch. This could be accredited to the incapacity of fat to retain moisture during cooking compared to the starch, which is in line with the water retention property difference of the patties stated in Table 1 [35]. The pH values of the patties were stable throughout the process, therefore, the use of starch did not affect the pH of the patties during storage or after cooking. The percent diameter reduction after cooking was 18.3 for the 5% fat content (control) and 23.5 for the 15% fat content, whereas for the Osan starch patties it was 7.7 and 8.6%, while the native starch exhibited reduction as 13.4 and 16.45%, respectively. This shows significantly lower diameter reduction of Osan starch patties compared to the control and the native starch. Thereby, the stability of the patties network structure due to Osan starch swelling and the formation of a semi solid gel that stabilized the diameter, is evident (Table 1). This can be attributed to the amphiphilic (hydrophilic-hydrophobic nature) property of the Osan starch. The control exhibited the greatest diameter loss after cooking by virtue of increased animal fat content. Park et al. [36] reported diameter and thickness reduction for pork patties decreased with increase in the content of black rice powder (rice powder is about 75% starch). The dimensional shrinkage followed the same trend as the diameter reduction. Consequently, the score of the cooking loss, moisture retention, diameter reduction, and dimensional reduction favored Osan over native starch. Cornejo-Ramírez et al. [37] reported that water absorption, swelling power, and viscosity of Osan starch is superior to native starch. The high moisture retention of Osan starch lead to highly viscous gel with emulsifying power capable of holding fat and water and form a gel with little un-noticeable pores (space within the structure) and improve the sensory characteristics of the patties [33]. On the other hand, the control beef patties exhibited poor binding, limited protein network structure, or entrapments of ingredients as shown by the sizable pores in the protein network. These results are in agreement with [38–41]. The thickness of the control increased at higher fat content, but with native starch, significantly lower thickness was observed at high starch content. Osan starch patties exhibited significantly higher thickness compared to the control and the native starch, since the thickness was almost twice as much (Table 1).

Table 1. The effect of 5 and 15% native, octenyl-succinylated (Osan) corn starch, and 5% animal fat on the physiochemical characteristics of beef patties.

	%					
	Control [7]		Native Starch		Osan Starch	
	5%	15%	5%	15%	5%	15%
M.C.U [2]	69.81 ± 0.72 [1]	63.72 ± 0.25 [b,1]	70.04 ± 0.12 [a]	66.25 ± 0.40 [a]	70.47 ± 0.31 [a]	64.41 ± 0.50 [b]
M.C.C [3]	42.45 ± 0.15 [b]	51.19 ± 0.10 [b]	60.69 ± 0.20 [a]	56.56 ± 0.83 [a]	61.14 ± 0.34 [a]	56.90 ± 0.32 [a]
C. loss	41.22 ± 0.18 [a]	45.29 ± 0.50 [a]	19.53 ± 0.11 [b]	09.50 ± 0.38 [b]	18.27 ± 0.20 [c]	08.45 ± 0.58 [b]
M.R [4]	35.82 ± 0.53 [c]	43.83 ± 0.44 [b]	69.96 ± 0.58 [b]	77.37 ± 1.31 [a]	71.20 ± 0.58 [a]	79.50 ± 0.80 [a]
Diameter [5]	18.37 ± 0.27 [a]	23.45 ± 1.08 [a]	13.36 ± 1.01 [b]	16.44 ± 2.05 [b]	07.71 ± 0.38 [c]	08.63 ± 2.51 [c]
D.S [6]	24.02 ± 0.19 [a]	28.98 ± 0.14 [a]	17.87 ± 0.26 [b]	18.02 ± 0.07 [b]	15.86 ± 0.13 [c]	10.81 ± 0.16 [c]
Thickness [8]	06.32 ± 0.49 [c]	14.51 ± 0.20 [b]	14.02 ± 1.57 [b]	07.09 ± 0.95 [c]	25.05 ± 0.28 [a]	18.10 ± 0.00 [a]
Firmness	121.97 ± 7.72 [a]	88.89 ± 4.25 [a]	43.56 ± 1.19 [b]	78.51 ± 2.08 [b]	40.96 ± 1.70 [b]	71.62 ± 1.54 [c]
pH	5.55 ± 0.01 [b]	5.53 ± 0.00 [a]	5.66 ± 0.06 [a]	5.60 ± 0.02 [b]	5.69 ± 0.02 [a]	5.62 ± 0.03 [a]

[1] The statistical analysis was done separately for the 5% and for the 15%; [2] M.C.U = moisture content of uncooked patties. [3] M.C.C = moisture content of cooked patties. [4] M.R = moisture retention; [5] diameter reduction; [6] D.S = dimensional shrinkage. [7] control = animal fat; [8] thickness increase, [a–c] Values followed by different letters within each row are significantly different ($p \leq 0.05$).

3.3.2. Effect of Storage on the Cooking Properties

The effect of storage on cooking loss, firmness, and other cooking properties is presented in Table 2. Although longer storage time significantly increased the tested parameter of patties containing 5% starch or animal fat, Osan starch performed better than the control or the native starch under the same storage conditions. The storability of patties significantly enhanced when 15% of starch or animal fat were added. Osan starch improved all tested parameters at 15% addition compared to 5%. The thickness of the control remained the same at 5% addition regardless of the storage time, but either of the starches reduced the thickness at longer storage time, significantly (Table 2). Nonetheless, Osan starch increased the thickness, significantly, compared to the control or the native starch especially after 0 or 30-day storage (Table 2). The same trend was observed for the 15% addition, where the thickness decreased after longer storage time for all samples, but the drop was less for Osan starch (Table 3). The firmness of all three patties increased after longer storage time regardless of the added fat or starch level. The increase in the firmness after longer storage time can be attributed to the moisture loss, however, Osan starch exhibited the least increase in firmness. Yang et al. [42] reported the addition of 4% modified waxy maize starch to low-fat frankfurters leading to reduced moisture loss by up to 7.2%. However, Claus and Hunt [43] also reported that, modified waxy maize starch applied to low-fat bologna was more effective in controlling moisture losses relative to native wheat starch, which is consistent with a very low retrogradation of waxy starch [34,35]. There are conflicting reports regarding the effect of starch on the cooking loss of meat products, especially for low muscle products such as frankfurters and bologna [33]. Cooking loss increase was observed for all samples at longer storage time, but Osan cooking loss was much less than the other treatments (Tables 2 and 3).

Table 2. Effect of 5% octenyl-succinylated (Osan) corn starch, native starch, and animal fat on the physiochemical characteristics of meat patties after storage at −20 °C for 0, 30, and 60 days.

Parameter		Days		
		0	30	60
Cooking loss	Control [6]	41.22 ± 0.18 [c,1]	51.85 ± 0.50 [b]	57.80 ± 0.62 [a]
	Native	19.53 ± 0.11 [c]	28.17 ± 1.35 [b]	36.10 ± 0.14 [a]
	Osan	18.27 ± 0.20 [c]	27.39 ± 0.23 [b]	33.58 ± 1.33 [a]
Diameter [2]	Control	18.37 ± 0.27 [b]	20.71 ± 0.58 [a]	21.70 ± 0.54 [a]
	Native	13.36 ± 1.01 [b]	15.62 ± 0.04 [a]	16.66 ± 0.14 [a]
	Osan	07.71 ± 0.38 [b]	09.99 ± 1.66 [a,b]	11.75 ± 0.75 [a]
D. S [3]	Control	24.02 ± 0.19 [c]	26.08 ± 0.08 [b]	27.79 ± 0.21 [a]
	Native	17.87 ± 0.26 [b]	21.25 ± 0.16 [a]	22.36 ± 1.07 [a]
	Osan	15.86 ± 0.13 [b]	16.38 ± 0.38 [a,b]	16.54 ± 0.48 [a]
Thickness [4]	Control	06.32 ± 0.49 [a]	05.80 ± 0.45 [a]	05.61 ± 0.98 [a]
	Native	14.02 ± 1.57 [a]	08.25 ± 0.07 [b]	06.13 ± 0.47 [b]
	Osan	25.05 ± 0.28 [a]	12.20 ± 0.67 [b]	07.47 ± 0.00 [c]
Firmness [5]	Control	121.97 ± 7.72 [c]	209.57 ± 4.05 [b]	307.19 ± 0.81 [a]
	Native	43.56 ± 1.19 [c]	92.93 ± 7.40 [b]	127.84 ± 8.59 [a]
	Osan	40.96 ± 1.70 [c]	82.73 ± 0.05 [b]	112.61 ± 3.84 [a]
pH	Control	05.63 ± 0.07 [a]	05.39 ± 0.05 [b]	05.35 ± 0.02 [b]
	Native	5.66 ± 0.06 [a]	5.39 ± 0.00 [b]	5.34 ± 0.02 [b]
	Osan	05.69 ± 0.02 [a]	05.33 ± 0.02 [b]	05.36 ± 0.01 [b]

[1] Values followed by different letters ([a–c]) within each row are significantly different ($p \leq 0.05$); [2] diameter reduction; [3] D.S = dimensional shrinkage; [4] thickness increase; [5] firmness in Newton; [6] control = animal fat.

Table 3. Effect of 15% octenyl-succinylated (Osan) corn starch, native starch, and animal fat on the physiochemical characteristics of meat patties after storage at −20 °C for 0, 30, and 60 days.

Parameter		Days		
		0	30	60
Cooking loss	Control [6]	45.29 ± 0.50 [b,1]	53.43 ± 1.65 [a]	55.08 ± 0.67 [a]
	Native	09.50 ± 0.38 [c]	20.45 ± 1.97 [b]	23.67 ± 1.32 [a]
	Osan	08.45 ± 0.58 [c]	17.86 ± 0.15 [b]	21.15 ± 1.39 [a]
Diameter [2]	Control	22.63 ± 0.18 [a]	23.23 ± 1.33 [a]	24.51 ± 0.31 [a]
	Native	14.03 ± 0.13 [c]	16.63 ± 0.04 [b]	18.68 ± 0.71 [a]
	Osan	05.43 ± 0.30 [b]	09.59 ± 0.06 [a]	10.87 ± 0.97 [a]
D. S [3]	Control	28.98 ± 0.14 [a]	29.93 ± 1.96 [a]	30.60 ± 0.30 [a]
	Native	18.02 ± 0.07 [b]	20.85 ± 1.22 [a]	22.53 ± 0.99 [a]
	Osan	10.81 ± 0.16 [b]	14.84 ± 0.50 [a]	15.33 ± 1.10 [a]
Thickness [4]	Control	14.51 ± 0.20 [a]	07.45 ± 0.70 [b]	02.72 ± 0.00 [c]
	Native	07.09 ± 0.95 [a]	05.21 ± 0.09 [a,b]	04.12 ± 0.45 [b]
	Osan	18.10 ± 0.00 [a]	14.39 ± 0.60 [b]	10.18 ± 0.00 [c]
Firmness [5]	Control	88.89 ± 4.25 [c]	163.63 ± 9.22 [b]	278.68 ± 14.65 [a]
	Native	78.50 ± 2.08 [b]	137.84 ± 6.30 [a]	158.23 ± 16.32 [a]
	Osan	71.62 ± 1.54 [c]	100.37 ± 8.98 [b]	127.81 ± 8.63 [a]
pH	Control	05.53 ± 0.00 [a]	05.43 ± 0.00 [b]	05.19 ± 0.00 [c]
	Native	05.60 ± 0.02 [a]	05.32 ± 0.00 [b]	05.07 ± 0.01 [c]
	Osan	05.58 ± 0.04 [a]	05.38 ± 0.05 [b]	05.12 ± 0.05 [c]

[1] Values followed by different letters ([a–c]) within each row are significantly different ($p \leq 0.05$); [2] diameter reduction; [3] D.S = dimensional shrinkage; [4] thickness increase; [5] firmness in Newton; [6] control = animal fat.

3.3.3. Sensory Attributes of Cooked Beef Meat Patties

The sensory evaluation results are presented in Table 4. The incorporation of native or Osan starches at 5% did not have any significant effect on the juiciness, flavor, color, or acceptability of the patties, but the tenderness was improved significantly ($p \leq 0.05$). The addition of 15% of either starches significantly improved the juiciness, flavor and, the overall acceptability of the patties, whereas the tenderness and the color were slightly improved more with Osan starch. After storage, the sensory evaluation showed superior performance of Osan over the control and the native starch, whereas longer storage appeared to have negative effect on the parameter of samples containing native starch (Table 5). Once again, samples with 15% Osan starch scored higher than those with 5%. Nonetheless, storage at −20 °C for 60 days appeared to have limited effect on the overall acceptability of the patties, but Osan starch patties scored higher. Higher starch content facilitates stable protein-starch matrices, where hydrogen and covalent bonding and charge-charge interactions occur [44].

Table 4. The effect of different levels of octenyl-succinylated (Osan) corn starch, native starch, and animal fat on the sensory characteristics of meat patties. The test was based on a 9 points hedonic.

Sensory Characteristics	Control [2]		Native Starch		Osan Starch	
	5%	15%	5%	15%	5%	15%
Tenderness	7.30 ± 0.26 [a,b,1]	6.43 ± 1.08 [b,1]	7.53 ± 1.12 [b]	7.53 ± 1.05 [a]	8.30 ± 0.65 [a]	8.03 ± 0.87 [a]
Juiciness	6.53 ± 1.26 [a]	5.96 ± 0.82 [c]	7.07 ± 1.18 [a]	6.80 ± 1.21 [b]	7.53 ± 1.26 [a]	8.23 ± 0.83 [a]
Flavor	7.23 ± 1.36 [a]	6.61 ± 1.19 [b]	7.53 ± 1.19 [a]	7.23 ± 1.23 [a,b]	7.84 ± 0.98 [a]	7.89 ± 0.95 [a]
Color	7.92 ± 0.95 [a]	7.48 ± 1.12 [a]	7.38 ± 1.26 [a]	7.59 ± 1.02 [a]	7.84 ± 0.80 [a]	7.92 ± 0.93 [a]
Acceptability	6.42 ± 1.22 [a]	6.50 ± 1.13 [b]	7.50 ± 1.32 [a]	7.61 ± 1.24 [a]	7.61 ± 1.30 [a]	7.78 ± 1.08 [a]

[1] The statistical analysis was done separately for the 5% and for the 15%; [(a–c)] values followed by different letters within each row and addition percent are significantly different ($p \leq 0.05$); [2] control = animal fat; the statistical analysis was done for the 5%, separate from the 15%.

Table 5. Effects of different levels of octenyl succinylated (Osan) corn starch, native starch, and animal fat on the sensory characteristics of meat patties after storage at −20 for different days.

		Days					
		0		30		60	
Parameter		5%	15%	5%	15%	5%	15%
Tenderness	Control [1]	7.30 ± 0.94 [a,1]	6.43 ± 1.08 [a,1]	6.30 ± 1.31 [b]	6.38 ± 1.19 [a]	6.53 ± 0.96 [a,b]	6.46 ± 0.96 [a]
	Native	7.53 ± 1.12 [a]	7.53 ± 1.05 [a,b]	7.46 ± 1.26 [a]	7.84 ± 0.98 [a]	7.15 ± 1.28 [a]	6.65 ± 1.12 [b]
	Osan	8.38 ± 0.65 [a]	8.03 ± 0.87 [a]	7.92 ± 1.18 [a,b]	8.07 ± 1.03 [a]	7.46 ± 0.87 [b]	7.84 ± 1.28 [a]
Juiciness	Control	6.38 ± 1.12 [a]	5.96 ± 0.82 [a]	6.38 ± 1.04 [a]	6.34 ± 1.06 [a]	6.46 ± 0.96 [a]	5.92 ± 0.86 [a]
	Native	7.07 ± 1.18 [a]	6.80 ± 1.2 [b]	7.84 ± 0.84 [a]	8.00 ± 0.81 [a]	6.92 ± 1.25 [a]	6.94 ± 1.19 [a,b]
	Osan	7.53 ± 1.26 [a]	8.23 ± 0.83 [a]	7.53 ± 1.05 [a]	7.53 ± 1.05 [a]	7.53 ± 0.66 [a]	7.38 ± 0.96 [a]
Flavor	Control	7.46 ± 1.12 [a]	6.61 ± 1.19 [a]	7.00 ± 1.29 [a]	6.92 ± 1.03 [a]	6.69 ± 1.10 [a]	9.92 ± 1.18 [a]
	Native	7.53 ± 1.19 [a]	7.23 ± 1.23 [a]	8.16 ± 0.83 [a]	7.53 ± 0.96 [a]	7.53 ± 0.96 [a]	6.92 ± 1.32 [a]
	Osan	7.84 ± 0.98 [a]	7.89 ± 0.95 [a]	7.00 ± 1.15 [a]	7.46 ± 1.05 [a]	7.07 ± 1.32 [a]	7.00 ± 0.91 [a]
Color	Control	7.92 ± 0.95 [a]	7.48 ± 1.12 [a]	7.23 ± 1.36 [a]	7.23 ± 1.16 [a]	7.15 ± 1.28 [a]	7.46 ± 1.05 [a]
	Native	7.38 ± 1.26 [a]	7.59 ± 1.02 [a]	7.84 ± 1.14 [a]	7.92 ± 0.95 [a]	7.69 ± 1.18 [a]	7.34 ± 1.14 [a]
	Osan	7.84 ± 0.80 [a]	7.92 ± 0.93 [a]	7.69 ± 0.94 [a]	8.23 ± 0.72 [a]	7.92 ± 1.11 [a]	7.42 ± 1.18 [a]
Acceptability	Control	6.77 ± 0.95 [a]	6.50 ± 1.13 [a]	6.65 ± 1.23 [a]	6.95 ± 1.09 [a]	6.76 ± 1.05 [a]	6.81 ± 1.06 [a]
	Native	7.50 ± 1.32 [a]	7.61 ± 1.24 [a]	7.64 ± 0.96 [a]	7.73 ± 1.06 [a]	7.12 ± 1.20 [a]	7.18 ± 0.89 [a]
	Osan	7.73 ± 1.11 [a]	7.78 ± 1.08 [a]	7.69 ± 1.02 [a]	7.63 ± 1.20 [a]	7.45 ± 0.81 [a]	7.20 ± 1.04 [a]

[1] Control = animal fat; [(a–c)] values followed by different letters within each row and addition percent are significantly different ($p \leq 0.05$); the statistical analysis was done for the 5% separate from the 15%.

3.3.4. Color of Raw and Cooked Patties

Consumer acceptability of meat products is dependent on its color because it is indicative of freshness. The effect of starch on the surface color of beef patties is presented in Table 6. The control sample exhibited the lowest redness (a*) values compared to the native and Osan starches, regardless of the added amount, but Osan starch had higher a* value. The incorporation of plant-based material was reported to increase the a* value of cooked meat [45]. Higher a* values of Osan starch-containing patties compared to control and the native starch indicate color stabilization, since the reduction in a* values is suggestive of myoglobin oxidation and the formation of net myoglobin [46]. Other researchers reported increase in meat redness after incorporation of potato starches, whereas cassava starch reduced the a*. Therefore, the type of starch as well as the amount of the added starch can be considered as factors that affect beef patties redness. This is obvious on the magnitude of the effect of native or Osan starch on the a* of the patties (Table 6). The amount of the incorporated Osan (5 or 15%) had slight effect on a* (Table 6). The redness of the control increased as a function of the added amount, but it increased significantly after storage for 60 days (Table 7). Sample with Osan had the highest a* as a function of storage time (60 days). The lightness (l*) of the samples and the control stayed almost the same at 5% incorporation, but at 15% Osan starch exhibited significantly higher l* value. The higher lightness could be attributed to the dilution of the color by the added starch. Reports in the literature mentioned reduction or stability of the l* based on the incorporated material into the patties [47]. No reduction in lightness was observed after storage for 30 or 60 days for either 5 or 15% incorporation. The yellowness (b*) of the control increased compared to the starch-containing samples where Osan exhibited the most b* value, but after storage for 60 days a drop in b* values was observed (Table 7).

Table 6. The effect of octenyl-succinylated (Osan) corn starch, native starch, and animal fat on the color characteristics of un-cooked meat patties after 6 h.

	Control [2]		Native Starch		Osan Starch	
	5%	15%	5%	15%	5%	15%
L* [3]	36.96 ± 1.68 [a,1]	40.08 ± 0.32 [a,1]	33.17 ± 0.36 [b]	33.00 ± 0.71 [c]	32.47 ± 0.44 [b]	37.59 ± 0.31 [b]
a* [4]	3.43 ± 0.46 [a]	2.46 ± 0.19 [b]	2.68 ± 0.51 [b]	2.80 ± 0.16 [b]	3.62 ± 0.08 [a]	3.49 ± 0.10 [a]
b* [5]	14.40 ± 0.51 [a]	15.66 ± 0.07 [a]	10.55 ± 0.37 [c]	13.36 ± 0.12 [c]	10.18 ± 0.21 [b]	14.86 ± 0.23 [b]
a*/b*	0.24 ± 0.41	0.16 ± 0.08	0.25 ± 0.21	0.21 ± 0.51	0.36 ± 0.31	0.23 ± 0.11

[1] Values followed by different letters ([a–c]) within each row are significantly different ($p \leq 0.05$); [2] control = animal fat. [3] L* = lightness, [4] a* = redness and [5] b* = yellowness; the statistical analysis was done for the 5% separate from the 15%.

The color of the cooked patties 6 h after preparation is presented in Table 8. The redness (a*) of cooked patties showed no significant difference between the control at 5% addition and Osan, but Osan starch exhibited significantly higher a* at 15% addition, whereas native starch reduced the redness. The lightness (l*) and yellowness (b*) of cooked patties was reduced in the presence of native or Osan starches. The effect of storage time on the color of the cooked patties is listed in Table 9. The redness (a*) of the cooked control did not change significantly after storage at either of the fat additions, whereas Osan starch increased the redness of the cooked patties compared to the control or native starch through storage time, especially after 60 days. Samples exhibited an increase in lightness and yellowness after longer storage time regardless of the amount of the added starch or animal fat (Table 9). The ratio of a*/b* is often used to describe color quality. The redness of the control was significantly reduced by the addition of 15% animal fat compared to 5%, but native starch maintained similar color for both additions. The addition of 5% Osan starch significantly reduced the yellowness which is obvious on the a*/b* value (Table 6). Therefore, animal fat added at 15% and Osan starch at 5% had the most improvement on the color of the cooked patties, which could be interpreted as stabilizing effect of Osan starch. This led to the higher redness, which indicates myoglobin color stability. Reports in the literature showed how the addition of processed plant leaf materials can reduce the a*/b* of beef patties which means lower redness [48].

Table 7. The effect of octenyl-succinylated (Osan) corn starch, native starch, and animal fat on the color characteristics of un-cooked meat patties at −20 for different days.

		0		30		60	
		5%	15%	5%	15%	5%	15%
L* [3]	Control [2]	[1] 36.96 ± 1.68 [a]	[1] 40.08 ± 0.32 [a]	36.84 ± 0.52 [a]	34.76 ± 0.69 [c]	38.26 ± 0.30 [a]	36.29 ± 0.54 [b]
	Native	33.17 ± 0.36 [b]	33.00 ± 0.71 [b]	35.24 ± 0.20 [a]	36.85 ± 0.85 [a]	35.30 ± 0.52 [a]	37.18 ± 0.10 [a]
	Osan	32.47 ± 0.44 [b]	37.59 ± 0.31 [b]	37.82 ± 1.27 [a]	39.40 ± 0.69 [a]	38.14 ± 0.40 [a]	38.95 ± 0.56 [a,b]
a* [4]	Control	3.43 ± 0.18 [a]	2.46 ± 0.19 [b]	3.00 ± 0.18 [a]	3.28 ± 0.04 [a]	3.20 ± 0.19 [a]	3.39 ± 0.32 [a]
	Native	2.68 ± 0.46 [b]	2.80 ± 0.16 [b]	3.72 ± 0.33 [b]	4.88 ± 0.10 [a]	5.50 ± 0.45 [a]	4.49 ± 0.30 [a]
	Osan	3.62 ± 0.08 [b]	3.49 ± 0.10 [c]	3.72 ± 0.33 [b]	4.89 ± 0.30 [b]	4.62 ± 0.12 [a]	6.80 ± 0.77 [a]
b* [5]	Control	14.40 ± 0.51 [a]	15.66 ± 0.07 [a]	12.81 ± 0.89 [a]	13.21 ± 0.49 [b]	15.68 ± 0.16 [a]	13.57 ± 1.51 [b]
	Native	10.55 ± 0.37 [c]	13.36 ± 0.12 [b]	14.67 ± 0.25 [b]	14.90 ± 0.89 [a]	15.68 ± 0.09 [a]	16.15 ± 0.15 [a]
	Osan	10.18 ± 0.21 [c]	14.86 ± 0.23 [b]	14.67 ± 0.59 [b]	15.19 ± 0.57 [b]	15.69 ± 0.59 [a]	16.70 ± 0.15 [a]

[1] Values followed by different letters ([a–c]) within each row are significantly different ($p \leq 0.05$); [2] control = animal fat. [3] L* = lightness, [4] a* = redness and [5] b* = yellowness; the statistical analysis was done for the 5% separate from the 15%.

Table 8. The effect of octenyl-succinylated (Osan) corn starch, native starch, and animal fat on the color characteristics of cooked meat patties after 6 h.

	[2] Control		Native Starch		Osan Starch	
	5%	15%	5%	15%	5%	15%
L* [3]	[1] 31.46 ± 1.68 [a]	[1] 30.46 ± 0.32 [a]	32.17 ± 0.36 [b]	32.38 ± 0.71 [c]	31.49 ± 0.44 [b]	38.50 ± 0.31 [b]
a* [4]	9.52 ± 0.46 [a]	9.69 ± 0.19 [b]	10.30 ± 0.51 [b]	10.91 ± 0.16 [b]	11.42 ± 0.08 [a]	11.66 ± 0.10 [a]
b* [5]	8.72 ± 0.51 [a]	8.32 ± 0.07 [a]	9.65 ± 0.37 [b]	11.77 ± 0.12 [c]	9.96 ± 0.21 [b]	11.88 ± 0.23 [b]
a*/b*	1.09 ± 0.41	1.16 ± 0.08	1.01 ± 0.21	0.93 ± 0.51	1.15 ± 0.31	0.98 ± 0.11

[1] Values followed by different letters ([a–c]) within each row are significantly different ($p \leq 0.05$); [2] control = animal fat. [3] L* = lightness, [4] a* = redness and [5] b* = yellowness; the statistical analysis was done for the 5% separate from the 15%.

Table 9. The effect of octenyl-succinylated (Osan) corn starch, native starch and animal fat on the color characteristics of cooked meat patties at −20 for different days.

		0		30		60	
		5%	15%	5%	15%	5%	15%
L* [3]	Control [2]	31.46 ± 1.68 [b,1]	[1] 30.46 ± 0.32 [b]	32.45 ± 0.52 [b]	36.47 ± 0.69 [a]	34.89 ± 0.30 [a]	37.27 ± 0.54 [a]
	Native	32.86 ± 0.36 [b]	32.38 ± 0.71 [c]	40.82 ± 0.20 [a]	42.64 ± 0.85 [a]	36.41 ± 0.52 [b]	45.63 ± 0.10 [a]
	Osan	31.49 ± 0.44 [b]	38.50 ± 0.31 [b]	36.96 ± 1.27 [a]	45.52 ± 0.69 [a]	38.09 ± 0.40 [a]	46.43 ± 0.56 [a]
a* [4]	Control	8.27 ± 0.18 [a]	9.69 ± 0.19 [a]	8.79 ± 0.18 [a]	10.02 ± 0.04 [a]	9.06 ± 0.19 [a]	8.28 ± 0.32 [b]
	Native	9.65 ± 0.46 [b]	10.91 ± 0.16 [a]	13.38 ± 0.33 [a]	9.83 ± 0.10 [a]	12.22 ± 0.45 [a]	9.90 ± 0.30 [a,b]
	Osan	9.96 ± 0.08 [b]	11.66 ± 0.10 [a]	11.75 ± 0.33 [a,b]	11.80 ± 0.30 [a]	12.65 ± 0.12 [a]	10.67 ± 0.77 [a]
b* [5]	Control	9.52 ± 0.51 [a]	8.32 ± 0.07 [b]	8.64 ± 0.89 [a]	10.72 ± 0.49 [a]	7.91 ± 0.16 [a]	10.52 ± 1.51 [a]
	Native	10.30 ± 0.37 [a,b]	11.77 ± 0.12 [c]	9.83 ± 0.25 [b]	14.42 ± 0.89 [b]	10.76 ± 0.09 [a]	15.48 ± 0.15 [a]
	Osan	11.24 ± 0.21 [a]	11.88 ± 0.23 [a]	10.17 ± 0.59 [b]	15.40 ± 0.57 [a]	10.78 ± 0.59 [a,b]	15.43 ± 0.15 [a]

[1] Values followed by different letters ([a–c]) within each row are significantly different ($p \leq 0.05$); [2] control = animal fat. [3] L* = lightness, [4] a* = redness and [5] b* = yellowness; the statistical analysis was done for the 5% separate from the 15%.

4. Conclusions

In this study we examined the quality parameters of patties prepared from ground beef and processed with the addition of Osan starch consequently revealing varying positive changes in the quality properties of the final product. The addition of Osan starch did not alter the pH level or the organoleptic standards of the product. The most notable change was manifested by the increase in textural and microstructural properties and the significant improvement of the cooking characteristics such as: yield, moisture retention, patties redness, thickness, and decrease in cooking loss. Scanning electron microscope (SEM) images of the samples confirmed the rise of intermolecular interaction between the proteins and the Osan starch, which resulted in small pores on the surface of the patties.

Author Contributions: Conceptualization, A.A.M., M.S.A. and I.A.; methodology, M.F.E.O.; software, M.F.E.O.; validation, I.A.M.A., and A.A.M.; formal analysis, S.H. and M.I.I.; investigation, S.H.; resources, M.I.I.; data curation, A.A.Q.; writing—original draft preparation, A.A.M.; writing—review and editing, A.A.M.; visualization, S.H.; supervision, A.A.M.; project administration, A.A.M.; funding acquisition, A.A.Q. All authors have read and agreed to the published version of the manuscript.

Funding: Deanship of Scientific Research at King Saud University for funding this work through research group no RGP-114.

Institutional Review Board Statement: Not applicable.

Data Availability Statement: Data will be available upon request from the corresponding author.

Acknowledgments: The authors extend their appreciation to the Deanship of Scientific Research at King Saud University for funding this work through research group no RGP-114.

Conflicts of Interest: The authors declare no conflict of interest.

References

1. Udomkun, P.; Ilukor, J.; Mockshell, J.; Mujawamariya, G.; Okafor, C.; Bullock, R.; Nabahungu, N.L.; Vanlauwe, B. What are the key factors influencing consumers' preference and willingness to pay for meat products in Eastern DRC? *Food Sci. Nutr.* **2018**, *6*, 2321–2336. [CrossRef] [PubMed]
2. Domínguez, R.; Pateiro, M.; Agregán, R.; Lorenzo, J.M. Effect of the partial replacement of pork backfat by microencapsulated fish oil or mixed fish and olive oil on the quality of frankfurter type sausage. *J. Food Sci. Technol.* **2017**, *54*, 26–37. [CrossRef] [PubMed]
3. Wu, J.H.; Micha, R.; Mozaffarian, D. Dietary fats and cardiometabolic disease: Mechanisms and effects on risk factors and outcomes. *Nat. Rev. Cardiol.* **2019**, *16*, 581–601. [CrossRef] [PubMed]
4. Asioli, D.; Varela, P.; Hersleth, M.; Almli, V.L.; Olsen, N.V.; Naes, T. A discussion of recent methodologies for combining sensory and extrinsic product properties in consumer studies. *Food Qual. Prefer.* **2017**, *56*, 266–273. [CrossRef]
5. Li, K.; Liu, J.Y.; Fu, L.; Li, W.J.; Zhao, Y.Y.; Bai, Y.H.; Kang, Z.L. Effect of gellan gum on functional properties of low-fat chicken meat batters. *J. Texture Stud.* **2019**, *50*, 131–138. [CrossRef] [PubMed]
6. Yemenicioğlu, A.; Farris, S.; Turkyilmaz, M.; Gulec, S. A review of current and future food applications of natural hydrocolloids. *Int. J. Food Sci. Technol.* **2020**, *55*, 1389–1406. [CrossRef]
7. Akbari, M.; Eskandari, M.H.; Davoudi, Z. Application and functions of fat replacers in low-fat ice cream: A review. *Trends Food Sci. Technol.* **2019**, *86*, 34–40. [CrossRef]
8. Babu, A.S.; Parimalavalli, R.; Mohan, R.J. Effect of modified starch from sweet potato as a fat replacer on the quality of reduced fat ice creams. *J. Food Meas. Charact.* **2018**, *12*, 2426–2434. [CrossRef]
9. Alcázar-Alay, S.C.; Meireles, M.A.A. Physicochemical properties, modifications and applications of starches from different botanical sources. *Food Sci. Technol.* **2015**, *35*, 215–236. [CrossRef]
10. Barbut, S. Effects of regular and modified potato and corn starches on frankfurter type products prepared with vegetable oil. *Ital. J. Food Sci.* **2018**, *30*, 802–808.
11. Tesch, S.; Gerhards, C.; Schubert, H. Stabilization of emulsions by OSA starches. *J. Food Eng.* **2002**, *54*, 167–174. [CrossRef]
12. He, G.-Q.; Song, X.-Y.; Ruan, H.; Chen, F. Octenyl succinic anhydride modified early indica rice starches differing in amylose content. *J. Agric. Food Chem.* **2006**, *54*, 2775–2779. [CrossRef] [PubMed]
13. Bajaj, R.; Singh, N.; Kaur, A. Properties of octenyl succinic anhydride (OSA) modified starches and their application in low fat mayonnaise. *Int. J. Biol. Macromol.* **2019**, *131*, 147–157. [CrossRef] [PubMed]
14. Partheniadis, I.; Zarafidou, E.; Litinas, K.E.; Nikolakakis, I. Enteric Release Essential Oil Prepared by Co-Spray Drying Methacrylate/Polysaccharides—Influence of Starch Type. *Pharmaceutics* **2020**, *12*, 571. [CrossRef]
15. Rezler, R.; Krzywdzińska-Bartkowiak, M.; Piątek, M. The influence of the substitution of fat with modified starch on the quality of pork liver pâtés. *LWT* **2021**, *135*, 110264. [CrossRef]

16. Balic, R.; Miljkovic, T.; Ozsisli, B.; Simsek, S. Utilization of modified wheat and tapioca starches as fat replacements in bread formulation. *J. Food Process. Preserv.* **2017**, *41*, e12888. [CrossRef]
17. Han, J.-A.; BeMiller, J.N. Preparation and physical characteristics of slowly digesting modified food starches. *Carbohydr. Polym.* **2007**, *67*, 366–374. [CrossRef]
18. Alejandre, M.; Passarini, D.; Astiasarán, I.; Ansorena, D. The effect of low-fat beef patties formulated with a low-energy fat analogue enriched in long-chain polyunsaturated fatty acids on lipid oxidation and sensory attributes. *Meat Sci.* **2017**, *134*, 7–13. [CrossRef]
19. Hollenbeck, J.J.; Apple, J.K.; Yancey, J.W.; Johnson, T.M.; Kerns, K.N.; Young, A.N. Cooked color of precooked ground beef patties manufactured with mature bull trimmings. *Meat Sci.* **2019**, *148*, 41–49. [CrossRef]
20. AOAC. *Official Methods of Analysis*, 18th ed.; Association of Official Analytical Chemists: Washington, DC, USA, 2007.
21. El-Magoli, S.B.; Laroia, S.; Hansen, P. Flavor and texture characteristics of low fat ground beef patties formulated with whey protein concentrate. *Meat Sci.* **1996**, *42*, 179–193. [CrossRef]
22. Heydari, F.; Varidi, M.J.; Varidi, M.; Mohebbi, M. Study on quality characteristics of camel burger and evaluating its stability during frozen storage. *J. Food Meas. Charact.* **2016**, *10*, 148–155. [CrossRef]
23. Naveena, B.; Muthukumar, M.; Sen, A.; Babji, Y.; Murthy, T. Improvement of shelf-life of buffalo meat using lactic acid, clove oil and vitamin C during retail display. *Meat Sci.* **2006**, *74*, 409–415. [CrossRef] [PubMed]
24. Meilgaard, M.; Civille, G.; Carr, B. Selection and training of panel members. *Sens. Eval. Tech.* **1999**, *3*, 174–176.
25. Stone, H.; Sidel, J.L. Introduction to sensory evaluation. In *Sensory Evaluation Practices*, 3rd ed.; Academic Press: San Diego, CA, USA, 2004; pp. 1–19.
26. Luo, F.-X.; Huang, Q.; Fu, X.; Zhang, L.-X.; Yu, S.-J. Preparation and characterisation of crosslinked waxy potato starch. *Food Chem.* **2009**, *115*, 563–568. [CrossRef]
27. Nagaoka, S.; Tobata, H.; Takiguchi, Y.; Satoh, T.; Sakurai, T.; Takafuji, M.; Ihara, H. Characterization of cellulose microbeads prepared by a viscose-phase-separation method and their chemical modification with acid anhydride. *J. Appl. Polym. Sci.* **2005**, *97*, 149–157. [CrossRef]
28. Simsek, S.; Ovando-Martinez, M.; Marefati, A.; Sjöö, M.; Rayner, M. Chemical composition, digestibility and emulsification properties of octenyl succinic esters of various starches. *Food Res. Int.* **2015**, *75*, 41–49. [CrossRef]
29. Moreno, H.M.; Carballo, J.; Borderías, A.J. Use of microbial transglutaminase and sodium alginate in the preparation of restructured fish models using cold gelation: Effect of frozen storage. *Innov. Food Sci. Emerg. Technol.* **2010**, *11*, 394–400. [CrossRef]
30. Tseng, T.-F.; Liu, D.-C.; Chen, M.-T. Evaluation of transglutaminase on the quality of low-salt chicken meat-balls. *Meat Sci.* **2000**, *55*, 427–431. [CrossRef]
31. Kumar, Y.; Kairam, N.; Ahmad, T.; Yadav, D.N. Physico chemical, microstructural and sensory characteristics of low-fat meat emulsion containing aloe gel as potential fat replacer. *Int. J. Food Sci. Technol.* **2016**, *51*, 309–316. [CrossRef]
32. Rasaei, S.; Hosseini, S.E.; Salehifar, M.; Behmadi, H. Effect of modified starch on some physico-chemical and sensory properties of low fat Hamburger. *Iran. J. Vet. Med.* **2012**, *6*, 89–94.
33. Pietrasik, Z.; Soladoye, O. Use of native pea starches as an alternative to modified corn starch in low-fat bologna. *Meat Sci.* **2021**, *171*, 108283. [CrossRef] [PubMed]
34. Copeland, L.; Blazek, J.; Salman, H.; Tang, M.C. Form and functionality of starch. *Food Hydrocoll.* **2009**, *23*, 1527–1534. [CrossRef]
35. Joly, G.; Anderstein, B. Starches. In *Ingredients in Meat Products*; Springer: Berlin/Heidelberg, Germany, 2009; pp. 25–55.
36. Park, S.-Y.; Lee, J.-W.; Kim, G.-W.; Kim, H.-Y. Effect of black rice powder on the quality properties of pork patties. *Korean J. Food Sci. Anim. Resour.* **2017**, *37*, 71. [CrossRef]
37. Cornejo-Ramírez, Y.I.; Martínez-Cruz, O.; Del Toro-Sánchez, C.L.; Wong-Corral, F.J.; Borboa-Flores, J.; Cinco-Moroyoqui, F.J. The structural characteristics of starches and their functional properties. *CyTA-J. Food* **2018**, *16*, 1003–1017. [CrossRef]
38. Lin, K.; Keeton, J.; Gilchrist, C.; Cross, H. Comparisons of carboxymethyl cellulose with differing molecular features in low-fat frankfurters. *J. Food Sci.* **1988**, *53*, 1592–1595. [CrossRef]
39. Mittal, G.; Barbut, S. Effects of various cellulose gums on the quality parameters of low-fat breakfast sausages. *Meat Sci.* **1993**, *35*, 93–103. [CrossRef]
40. Ruusunen, M.; Vainionpää, J.; Puolanne, E.; Lyly, M.; Lähteenmäki, L.; Niemistö, M.; Ahvenainen, R. Physical and sensory properties of low-salt phosphate-free frankfurters composed with various ingredients. *Meat Sci.* **2003**, *63*, 9–16. [CrossRef]
41. Schuh, V.; Allard, K.; Herrmann, K.; Gibis, M.; Kohlus, R.; Weiss, J. Impact of carboxymethyl cellulose (CMC) and microcrystalline cellulose (MCC) on functional characteristics of emulsified sausages. *Meat Sci.* **2013**, *93*, 240–247. [CrossRef]
42. Yang, A.; Keeton, J.; Beilken, S.; Trout, G. Evaluation of some binders and fat substitutes in low-fat frankfurters. *J. Food Sci.* **2001**, *66*, 1039–1046. [CrossRef]
43. Claus, J.; Hunt, M. Low-fat, high added-water bologna formulated with texture-modifying ingredients. *J. Food Sci.* **1991**, *56*, 643–647. [CrossRef]
44. Schmitt, C.; Sanchez, C.; Desobry-Banon, S.; Hardy, J. Structure and technofunctional properties of protein-polysaccharide complexes: A review. *Crit. Rev. Food Sci. Nutr.* **1998**, *38*, 689–753. [CrossRef] [PubMed]
45. Shah, M.A.; Bosco, S.J.D.; Mir, S.A. Plant extracts as natural antioxidants in meat and meat products. *Meat Sci.* **2014**, *98*, 21–33. [CrossRef] [PubMed]
46. Mancini, R.; Hunt, M. Current research in meat color. *Meat Sci.* **2005**, *71*, 100–121. [CrossRef] [PubMed]

47. Muthukumar, M.; Naveena, B.; Vaithiyanathan, S.; Sen, A.; Sureshkumar, K. Effect of incorporation of *Moringa oleifera* leaves extract on quality of ground pork patties. *J. Food Sci. Technol.* **2014**, *51*, 3172–3180. [CrossRef]
48. Al-Juhaimi, F.; Ghafoor, K.; Hawashin, M.D.; Alsawmahi, O.N.; Babiker, E.E. Effects of different levels of *Moringa* (*Moringa oleifera*) seed flour on quality attributes of beef burgers. *CyTA-J. Food* **2016**, *14*, 1–9. [CrossRef]

Article

Low-Voltage Electrical Stimulation of Beef Carcasses Slows Carcass Chilling Rate and Improves Steak Color

Christina Bakker, Keith Underwood, Judson Kyle Grubbs * and Amanda Blair

Department of Animal Science, South Dakota State University, Brookings, SD 57007, USA; Christina.Bakker@sdstate.edu (C.B.); keith.underwood@sdstate.edu (K.U.); amanda.blair@sdstate.edu (A.B.)
* Correspondence: judson.grubbs@sdstate.edu

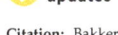

Citation: Bakker, C.; Underwood, K.; Grubbs, J.K.; Blair, A. Low-Voltage Electrical Stimulation of Beef Carcasses Slows Carcass Chilling Rate and Improves Steak Color. *Foods* **2021**, *10*, 1065. https://doi.org/10.3390/foods10051065

Academic Editor: Paulo Eduardo Sichetti Munekata

Received: 1 April 2021
Accepted: 8 May 2021
Published: 12 May 2021

Publisher's Note: MDPI stays neutral with regard to jurisdictional claims in published maps and institutional affiliations.

Copyright: © 2021 by the authors. Licensee MDPI, Basel, Switzerland. This article is an open access article distributed under the terms and conditions of the Creative Commons Attribution (CC BY) license (https://creativecommons.org/licenses/by/4.0/).

Abstract: Electrical stimulation (ES) is used in beef slaughter plants to improve tenderness; however, varying levels of low-voltage ES have not been well characterized. The objective was to evaluate the influence of two levels of low-voltage ES on temperature decline, pH, glycolytic potential, and meat quality. Forty-two beef carcasses were chosen from a commercial packing facility. One side of each carcass received either 40 or 80 volts of ES for 60 s at 45 min postmortem. The paired side of each carcass did not receive ES (Control). Temperature loggers were placed in the sirloin of 12 carcasses to record temperature decline. Longissimus muscle pH was measured at 1, 12, and 24 h, and 3 d postmortem. Strip steaks were fabricated for determination of meat quality. A treatment by time interaction was observed for carcass temperature decline ($p < 0.001$) where ES sides stayed warmer longer than Control sides. A treatment by time interaction was observed for pH decline with Control sides exhibiting an increased pH at 1 h postmortem ($p < 0.001$). Instrumental color values were increased for ES compared to Control sides ($p < 0.001$). These results indicate ES slows carcass temperature decline, hastens initial pH decline, and improves instrumental color. Similar results were observed between the ES treatments indicating either ES level may be used to achieve similar quality characteristics.

Keywords: beef; electrical stimulation; glycolytic potential; quality; temperature decline

1. Introduction

Electrical stimulation (ES) is a postmortem intervention utilized to enhance beef quality traits including color, tenderness, and flavor. Electrical stimulation is proposed to improve tenderness by reducing cold shortening [1], disrupting muscle structure [2], and increasing proteolytic activity [3]. Extra low-voltage ES is used on beef carcasses to facilitate the removal of blood from carcasses shortly after exsanguination, while low- and high-voltage ES is used to improve the tenderness and color of beef [4–6]. However, there are discrepancies among reports regarding the influence of varying levels of ES on beef quality traits. In a review by Adeyemi and Sazili [7], these discrepancies caused by varying levels of ES on beef quality are highlighted, with some authors reporting positive effects including improvements in tenderness and lean maturity, some reporting negative effects such as reduced color stability and water holding capacity, and others reporting no effect of ES on meat quality, thus concluding the need to further study the effective application of this technology. Throughout the beef industry in the United States, few plants utilize ES in the same manner. Some plants utilize extra low-voltage ES to facilitate blood removal, others apply low- or high-voltage ES to improve tenderness and lean maturity scores, some apply different ES voltage levels throughout the slaughter process, and yet others do not use ES at all. Thus, additional research is necessary to optimize ES applications to ensure beneficial effects are captured and deleterious effects are minimized. Therefore, the objective of this study was to evaluate the influence of two levels of low-voltage electrical stimulation applied at 45 min postmortem on temperature decline, muscle pH, instrumental color, glycolytic potential, and instrumental tenderness. We hypothesized the

ES treatments would increase carcass temperature, decrease muscle pH, increase glycolytic potential, improve tenderness, and increase instrumental L* and a* values compared to the non-stimulated sides, with the 80 V ES treatment having a greater impact on these traits than the 40 V treatment.

2. Materials and Methods

2.1. Carcass Selection and Electrical Stimulation Treatments

Cattle were shipped from feedlots to a commercial slaughter facility and held in lairage following normal plant operating guidelines and United States Department of Agriculture Food Safety Inspection Service regulations for beef slaughter. Source and history of the cattle is unknown. Carcasses (n = 42) were selected for comparison in this study. Three collections were conducted throughout the course of the production day (11 carcasses at 0900 h, 16 carcasses at 1200 h, and 15 carcasses at 1500 h). Carcasses were harvested using standard industry methods. Prior to chilling, paired sides were identified to compare the influence of 2 levels of ES. The left side of the carcasses were subjected to one of two ES treatments, (1) 80 V (ES80; n = 20) and (2) 40 V (ES40; n = 22), 45 min after exsanguination. The right side of each carcass was used as an unstimulated control. For both ES40 and ES80 treatments, the ES was administered through the carcass trolly as it moved over a section of electrically charged rail. Electrical stimulation was applied over a 60 s period where the carcasses received a 4 s pulse of electricity with approximately 2 s between each pulse. The remaining side of each carcass served as a negative control and did not receive ES (Control; n = 42).

2.2. Carcass Temperature and pH

Following application of ES treatments, all carcasses were placed on the same rail in a cooler set to hold at approximately 3 °C for 48 h. Carcass temperature decline was monitored from the timepoint the carcasses entered the cooler on paired sides by inserting a temperature probe (Temprecord Multitrip, Sensitech Inc. Beverly, MA, USA) into the sirloin of both sides of the first 4 carcasses selected at each of the 3 collection time points. Once the cooler was filled with carcasses, the spray chill system was activated to spray water for 1 min every 15 min for 24 h. Upon completion of the 48-h holding period, carcasses were ribbed and allowed to bloom for approximately 30 min before standard carcass data were collected. Longissimus muscle pH was recorded on the medial side of the muscle at the 12th rib position at 1 h postmortem, at the 11th rib at 12 h postmortem, and at the 10th rib at 24 h postmortem to establish a pH decline through the completion of rigor mortis. The pH recordings were taken at different locations on the muscle to avoid influence on pH by utilizing the same probe site.

2.3. Carcass Characteristics and Sample Collection

Fat thickness at the 12th rib (BF) and ribeye area (REA) were measured on both sides of each carcass by South Dakota State University (SDSU) personnel. Fat thickness at the 12th rib and REA measurements of the two sides were averaged and used to calculate USDA Yield Grade. Hot carcass weight (HCW) was recorded from each side and added together for a total hot carcass weight for the carcass. Boneless striploins (IMPS #180) were collected, transported under refrigerated conditions to SDSU, and fabricated into 2.54 cm steaks. Steaks were fabricated in a set order. The first anterior steak was immediately frozen, 3 d postmortem, and used for glycolytic potential (GP) analysis. The second through fourth steaks were aged for 3, 7, or 14 d, respectively, and utilized for Warner–Bratzler shear force (WBSF) and cook loss determination. The second anterior steak was also used to measure ultimate pH, 3 d postmortem. The seventh steak was used to evaluate instrumental lean color for each loin.

2.4. Glycolytic Potential

Glycolytic potential was determined as described by McKeith et al. [8] with minor modifications. Briefly, steaks designated for GP analysis were minced, snap frozen in liquid nitrogen and powdered using a Waring commercial blender (Model 51BL32, Waring Products Division, New Hartford, CT, USA) to produce a homogenous sample. Three g of powdered sample was weighed into a 50 mL plastic conical tube, allowed to thaw, and then homogenized for 75 s in 0.6 N perchloric acid. Samples were then digested using amyloglucosidase and 5.4 N potassium hydroxide and incubated for 3 h, inverting the tubes every 20 min to mix. Upon completion of the incubation step, 3N perchloric acid was added and samples were centrifuged at 4.4 °C for 5 min at $10,000 \times g$. Supernatant was collected and stored for analysis. Glucose levels were determined using a glucose assay kit (Glucose (HK) Assay Kit GAHK20, Millipore-Sigma, St. Louis, MO, USA) and absorbance was read at 340 nm (SpectraMax 190, Molecular Devices, San Jose, CA, USA). Lactate levels were determined by adding NAD+ in a glycine buffer to sample aliquots to form NADH. Samples were then read at 340 nm (SpectraMax 190, Molecular Devices, San Jose, CA, USA). Glycolytic potential of each sample was then calculated with the following equation: GP = 2(Glucose absorbance * 111.882) + (Lactate absorbance * 173.22).

2.5. Warner–Bratzler Shear Force and Cook Loss

Steaks utilized for WBSF were thawed at approximately 4 °C for 24 h prior to cooking. Steaks were cooked on a clamshell grill (George Foreman Indoor/Outdoor Grill model GGR62, Lake Forest, IL, USA) to an internal peak temperature of 71 °C as indicated by a temperature probe inserted to the geometric center of the steak (Atkins AquaTuff NSF Series Model 351, Middlefield, CT, USA). Steaks were then stored at approximately 4 °C overnight. Four h prior to evaluating shear force values, steaks were placed at room temperature and allowed to equilibrate. Six cores were removed parallel to the direction of the muscle fibers and then sheared once using a Warner–Bratzler shear machine (G-R Electric Manufacturing Company, Manhattan, KS, USA) equipped with a BFG 500 N basic force gauge (Mecmesin Ltd., West Sussex, UK) and peak shear force was recorded for each core. An average shear force value was calculated and recorded for each steak.

Cook loss was determined on steaks designated for WBSF. Raw steak weight was recorded with a balance (MWP, Cas Corporation, Seoul, South Korea) and after cooking, steaks were allowed to equilibrate to room temperature and weighed again. Cook loss was determined using the following equation: cook loss % = ((raw weight − cooked weight)/raw weight) × 100.

2.6. Instrumental Color

Steaks designated for color determination were allowed to bloom for 30 min prior to evaluation. L*, a*, and b* values were recorded at two locations (medial portion of the steak and lateral portion of the steak) using a handheld colorimeter (Chroma Meta CR-410, Konica Minolta, Ramsey, NJ, USA) equipped with a 50 mm aperture, 0° viewing angle, 2° standard observer, pulsed xenon lamp light source, and calibrated with a white tile (L* = 97.38, a* = 0.06, b* = 1.82). Measurements were averaged between both locations for each steak.

2.7. Statistical Analysis

The experiment utilized both sides of 42 carcasses in a completely randomized design. The data analysis was conducted using the MIXED model of SAS software (SAS Institute Inc., Cary, NC, USA) with fixed effect of treatment, random effect of carcass, and Toeplitz covariate structure. Hot carcass weights for both sides of each carcass were added together, and REA and BF measurements were averaged between sides. As HCW from both sides are needed to calculate USDA yield grades, carcass data were analyzed by ES treatment with data reported as ES40 or ES80 treatments. Contrast statements were used to compare Control vs. ES40 and ES80 sides (Control vs. ES), and ES40 vs. ES80 (ES Level). Peak internal cooking temperature was used as a covariate for cook loss and WBSF data. Temperature decline, WBSF, cook loss, and pH were considered repeated measures. Interactions of treatment and time were evaluated where appropriate and are reported when significant. Significance was determined when $p < 0.05$.

3. Results

3.1. Carcass Characteristics

Carcass characteristics are reported in Table 1. Hot carcass weight did not differ between ES40 and ES80 treatments ($p = 0.7200$). No differences were observed in REA ($p = 0.6172$). Fat thickness measured at the 12th rib was similar between the two treatments ($p = 0.9482$). The lack of differences in HCW, REA, and BF contributed to the absence of differences in overall USDA yield grade ($p = 0.5000$). The absence of differences in carcass characteristics between ES treatments indicates that carcass characteristics likely did not impact carcass chilling or meat quality data.

Table 1. Least square means for carcass characteristics of carcasses subjected to 40 or 80 V of electrical stimulation for 60 s in 4 s on, 2 s off intervals prior to chilling.

Variable	Treatment [1]		SEM [3]	p-Value
	ES40	ES80		
Hot carcass weight, kg	427.25	424.30	8.16	0.7200
Ribeye area [4], cm^2	85.06	87.44	4.71	0.6172
12th rib fat thickness [4], cm	1.62	1.61	0.14	0.9482
USDA YG [5]	3.86	3.70	0.23	0.5000

[1] ES40 = 40 V of electrical stimulation, ES80 = 80 V of electrical stimulation. [3] Standard error of means. [4] Carcass data measured between the 12th and 13th rib according to USDA standards. [5] USDA Yield Grade.

3.2. Carcass Temperature and PH

An ES by chilling time interaction was observed for temperature decline ($p < 0.0001$; Figure 1). Sides treated with ES prior to chilling had similar temperatures to non stimulated sides at the onset of chilling. By 30 min of chilling, ES sides had increased temperatures compared to sides that did not receive ES, regardless of ES level. This difference persisted until 24 h postmortem when temperature data loggers were removed from the carcasses. No differences in temperature between ES treatments were observed at any time point ($p > 0.05$).

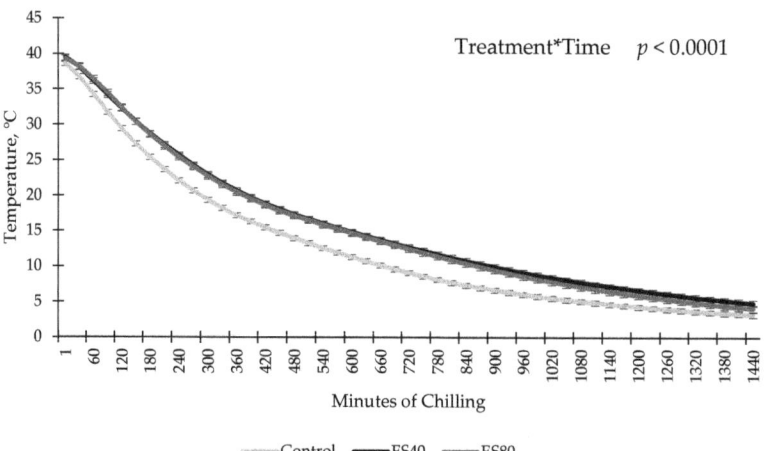

Figure 1. Temperature decline of carcass submitted to low-voltage electrical stimulation (ES) prior to chilling. Data are depicted as least square means ± SEM. Treatments are as follows: Control = no ES, ES40 = 40 V of ES, ES80 = 80 V of ES. Electrical stimulation was applied for 60 s in 4 s on, 2 s off intervals.

An ES treatment by chilling time interaction was observed for pH decline ($p < 0.0001$; Figure 2). At 1 h postmortem, the ES80 carcasses achieved the lowest pH, ES40 intermediate, and Control sustaining the highest pH value. The pH values measured at 12 and 24 h postmortem, as well as ultimate pH, did not differ among treatments ($p > 0.05$).

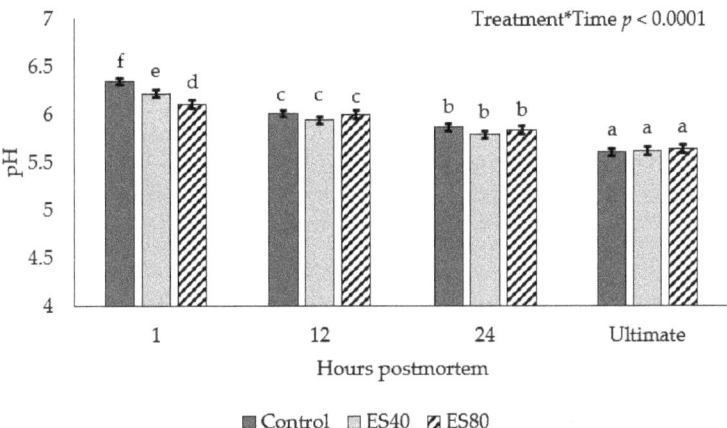

Figure 2. pH decline in beef carcasses subjected to low-voltage electrical stimulation prior to chilling. Data are depicted as least square means ± SEM. Treatments are as follows: Control = no ES, ES40 = 40 V of ES, ES80 = 80 V of ES. Electrical stimulation was applied for 60 s in 4 s on, 2 s off intervals. Measurements were taken at 1, 12, and 24 h postmortem in addition to ultimate pH. [a–f] Means with different subscripts differ ($p < 0.05$).

3.3. Glycolytic Potential

Glucose, lactate, and GP data are reported in Table 2. Glucose concentration did not differ between Control and ES sides ($p = 0.5825$) or between ES treatments ($p = 0.7308$). Additionally, no differences were observed between Control and ES sides ($p = 0.9557$) or between ES levels ($p = 0.5655$) for lactate concentration. Unsurprisingly, based on glucose and lactate results, GP did not differ between Control and ES sides ($p = 0.6760$), or between ES treatments ($p = 0.5784$).

Table 2. Warner–Bratzler shear force (WBSF), cook loss, glucose, lactate, and glycolytic potential (GP) of beef carcasses subjected to low-voltage electrical stimulation for 60 s in 4 s on, 2 s off intervals prior to chilling [1,2].

	Treatment [3]			Contrast p-Value	
Variable	Control	ES40	ES80	Control vs. ES [4]	ES Level [5]
Glucose, µmol/g	0.19 ± 0.006	0.19 ± 0.008	0.18 ± 0.008	0.5825	0.7308
Lactate, µmol/g	0.20 ± 0.005	0.20 ± 0.006	0.19 ± 0.006	0.9557	0.5655
GP, µmol/g	76.10 ± 1.66	76.02 ± 2.18	74.26 ± 2.28	0.6760	0.5784
WBSF, kg	3.84 ± 0.08	3.69 ± 0.10	3.64 ± 0.11	0.0220	0.7332
Cook loss, %	17.94 ± 0.28	18.34 ± 0.38	18.24 ± 0.39	0.3753	0.8536

[1] Least square means ± standard error of means. [2] No interaction was observed for aging day and electrical stimulation treatment (effect of aging day is reported in Table 3). [3] Carcasses subjected to 0 (Control), 40 (ES40), or 80 (ES80) V of electrical stimulation. [4] Control vs. ES contrast statement compares Control carcasses vs. 40 and 80 V treatments. [5] ES Level contrast statement compares 40 vs. 80 V treatments

Table 3. Least square means for Warner–Bratzler Shear Force (WBSF) and cook loss values of beef steaks aged 3, 7, or 14 d [1] (n = 42/day).

	Days Postmortem [1]			SEM [2]	p-Value
Variable	3	7	14		
WBSF, kg	3.70 [a]	3.84 [b]	3.63 [a]	0.08	0.0021
Cook loss, %	17.38 [a]	18.69 [b]	18.45 [b]	0.34	0.0127

[1] No interaction was observed for aging day and electrical stimulation treatment (effect of electrical stimulation treatment is reported in Table 2). [2] Standard error of means. [a,b] Means with different subscripts indicate a difference within row ($p < 0.05$).

3.4. Warner–Bratzler Shear Force and Cook Loss

Steaks from ES sides exhibited decreased shear force values compared to the Control sides ($p < 0.0220$; Table 2). However, when evaluating WBSF data between ES treatments, no differences were observed ($p = 0.7332$). Moreover, an aging day effect was observed for WBSF. Steaks aged 7 d had a greater shear force value compared to steaks aged for 3 or 14 d postmortem ($p = 0.0021$; Table 3).

The percentage of weight lost during cooking did not differ between Control and ES-treated sides ($p = 0.3753$; Table 2) nor were differences observed between sides treated with different ES levels ($p = 0.8536$). An aging day effect was observed for cook loss ($p = 0.0127$; Table 3) with steaks aged 3 d demonstrating less cook loss than steaks aged 7 or 14 d.

3.5. Instrumental Color

Steaks from ES sides were lighter ($p < 0.0001$; Table 4), redder ($p < 0.0001$; Table 4), and more yellow ($p < 0.0001$; Table 4) than control steaks. No differences were observed between ES treatments for lightness ($p = 0.4582$), redness ($p = 0.9460$), or yellowness ($p = 0.7079$).

Table 4. Instrumental color values of longissimus muscle from beef carcasses subjected to low-voltage electrical stimulation for 60 s in 4 s on, 2 s off intervals prior to chilling [1].

	Treatment [2]			Contrast p-Value	
Variable	Control	ES40	ES80	Control vs. ES [3]	ES Level [4]
L*	40.38 ± 0.34	42.28 ± 0.46	42.77 ± 0.48	<0.0001	0.4582
a*	24.94 ± 0.30	26.08 ± 0.33	26.06 ± 0.38	<0.0001	0.9460
b*	10.14 ± 0.27	11.30 ± 0.29	11.19 ± 0.34	<0.0001	0.7079

[1] Least square means ± standard error of means. [2] Carcasses subjected to 0 (Control), 40 (ES40), or 80 (ES80) V of electrical stimulation. [3] Control vs. ES contrast statement compares Control carcasses vs. 40 and 80 V treatments. [4] ES Level contrast statement compares 40 vs. 80 V treatments

4. Discussion

Previous research has shown temperature decline trends similar to the current data with ES reported to increase carcass temperature. Bowker et al. [9] measured the temperature decline of the longissimus dorsi in pigs electrically stimulated (six pulses, 60 Hz, 500 V, 1 s on and 2 s off) at 3 min postmortem, and observed an increase in temperature of ES-treated carcasses over the monitoring duration of 56 min. In both cases, the increase in temperature was likely caused by the heat generated by the muscle contractions caused by the ES treatment [9]. Conversely, Wiklund et al. [10] evaluated the temperature decline in the longissimus muscle of red deer carcasses stimulated with 90–95 V of ES for 55 s at the time of exsanguination, and found no differences compared to non-stimulated carcasses. Additionally, Kim et al. [11] evaluated the impact of low-voltage ES (100 V for 30 s) 90 min after exsanguination of beef carcasses, and also observed no differences in the temperature decline of the longissimus dorsi compared to non-stimulated sides. The conflicting results of Wiklund et al. [10] and Kim et al. [11] compared to the current study could be due to differences in species (beef vs. red deer) or time post exsanguination of the stimulation.

Electrical stimulation can cause an increase in the rate of postmortem muscle pH decline by increasing metabolic activity. McKenna et al. [12] observed differences in early pH measurements with ES sides showing decreased pH values compared to non-stimulated sides until 6 h postmortem when pH was similar, until cessation of pH measurements at 24 h postmortem. Moreover, Nichols and Cross [13] noted a similar trend in pH with ES sides displaying a rapid pH decline in the first 6 h postmortem. Kim et al. [11] noted a more dramatic decrease in longissimus muscle pH decline, with non-stimulated sides displaying an increased pH until 24 h postmortem. The rapid pH decline observed was likely caused by the increase in postmortem glycolysis, which resulted in a buildup of lactic acid in muscle at a faster rate than would occur without ES, but resulted in similar ultimate pH values [14,15].

Similar to the current study, Ding et al. [16] observed no differences in glucose or GP values for bison meat from carcasses stimulated with 400 V of ES compared to a non-stimulated control. Conversely, Ding et al. [16] did observe a difference in lactate concentrations; however, the samples were taken from carcasses prior to chilling and rigor mortis. The lack of differences in GP observed in this study was ideal as we could conclude the animals used in this study were at similar metabolic states prior to slaughter. Further, we can conclude that pre-harvest handling did not impact the ability of carcasses in this study to experience a normal rigor processes, such as pH decline, and the differences in pH observed in the present study were likely the result of the ES treatments.

There are several mechanisms by which ES is proposed to improve tenderness [17]. It has been reported that ES disrupts muscle structure at the Z-disk and I-band, causes formation of contraction nodes, and disrupts the integrity of the sarcoplasmic reticulum causing minor separation of myofibrils [2,18,19]. Electrical stimulation has also been proposed to inhibit cold shortening by preventing the temperature of the carcass from declining too rapidly [1,20]. Others hypothesize that improvements in tenderness following ES is caused by the activation of lysosomal enzymes and increased proteolysis while carcass temperature is still increased [2,3,10]. However, some studies have found little or no effects of ES on beef tenderness. Discrepancies between studies could be related to the level of voltage applied, duration of stimulation, or timing of ES after exsanguination [11,21]. However, most studies agree that tenderness development is a complex process that likely involves more than one of the previously discussed mechanisms [2,7,14].

It is unclear why steaks aged for 7 d had increased shear force values compared to steaks aged 3 d, when most normal aging curves would show a decrease in WBSF value as aging day increased during the first few weeks of aging. The steaks utilized for shear force were taken consecutively from the anterior end of the strip loin. Previous research suggests that steaks from those locations should have similar shear force values, likely eliminating the impact of steak location on tenderness [22,23]. The WBSF values for each aging period are below the established threshold for tenderness (4.6 kg) as perceived by consumers as outlined by Shackelford et al. [24]. Additionally, the difference among days is within the 0.5 kg of force described by Miller et al. [25] as the difference in shear force detectable by consumers preparing steak in their own home, indicating that the differences in shear force based on aging day are likely not detectable by the average consumer.

Cook loss was not impacted by treatment in the current study. These data are similar to the impact of high-voltage ES on cook loss of beef steaks [5,12] or bison steaks [16]. Additionally, Wiklund et al. [10] observed no effect of ES on drip loss of steaks from red deer. However, when evaluating treatment of beef carcasses with 100 V of ES at 1 h postmortem Savell et al. [26] observed increased cook loss for ES vs. control carcasses. However, cook loss in the current study was impacted by aging day. Similar results were observed by Shanks et al. [27] when evaluating cook loss over 35 d postmortem. Increases in cook loss over time may be the result of damage to cellular membranes, which would enable a greater amount of water to leak out of the muscle during cooking [27].

Steaks from ES sides in the current study were lighter, more red, and more yellow than steaks from Control sides. Similar results were observed in beef [5,6,26,28] and in bison [16].

The increased color values can be attributed to the increased oxygen permeability of the meat as a result of the damaged muscle fibers. Weakened muscle structure caused by the intense contractions that occurred during the ES treatment can allow oxygen to penetrate deeper into the muscles, resulting in a thicker layer of oxymyoglobin formation compared to non-stimulated carcasses [6,29,30].

5. Conclusions

The goal of this study was to evaluate the effects of varying levels of low-voltage ES on beef quality traits. Collectively, these results demonstrate that low-voltage ES can be an effective means to improve the tenderness and instrumental color scores of beef carcasses without increasing cook loss, potentially improving consumer satisfaction. Within this study the only differences observed between the ES40 and ES80 treatments were the early postmortem pH levels. Thus, beef processing facilities that implement low-voltage ES immediately before carcass chilling may be able to reduce the ES voltage levels to 40 V without detrimentally impacting the meat quality characteristics expected with increased voltages. Additionally, these data show that the desired appearance and palatability benefits of high-voltage ES may be attainable using decreased voltages.

Author Contributions: Conceptualization, C.B., J.K.G., K.U. and A.B.; methodology, C.B. and A.B.; validation, C.B., J.K.G., K.U. and A.B.; formal analysis, C.B. and A.B.; investigation, C.B., J.K.G., K.U. and A.B.; resources, J.K.G., K.U. and A.B.; data curation, C.B., J.K.G., K.U. and A.B.; writing—original draft preparation, C.B.; writing—review and editing, C.B., J.K.G., K.U. and A.B.; visualization, C.B. and A.B.; supervision, A.B.; project administration, A.B.; funding acquisition, J.K.G., K.U. and A.B. All authors have read and agreed to the published version of the manuscript.

Funding: This research was supported by state and federal funds appropriated to South Dakota State University including support from the U.S. Department of Agriculture National Institute of Food and Agriculture, through the Hatch Act (Accession number 1005460).

Data Availability Statement: The data presented in this study are available on request from the corresponding author. The data are not publicly available due to privacy restrictions.

Acknowledgments: The authors express their gratitude to DemKota Ranch Beef (Aberdeen, SD) for their assistance in data collection and support of this project.

Conflicts of Interest: The authors declare no conflict of interest. The funders had no role in the design of the study; in the collection, analyses, or interpretation of data; in the writing of the manuscript, or in the decision to publish the results.

References

1. Chrystall, B.B.; Hagyard, C. Electrical stimulation and lamb tenderness. *N. Z. J. Agric. Res.* **1976**, *19*, 7–11. [CrossRef]
2. Dutson, T.R.; Yates, D.L.; Smith, G.C.; Carpenter, Z.L.; Hostetler, R.L. Rigor onset before chilling. In Proceedings of the Reciprocal Meat Conference, Auburn, AL, USA, 12–15 June 1997; pp. 79–86.
3. Locker, R.H.; Daines, G.J. Tenderness in relation to the temperature of rigor onset in cold shortened beef. *J. Sci. Food Agric.* **1976**, *27*, 193–196. [CrossRef]
4. McKeith, F.K.; Smith, G.C.; Savell, J.W.; Dutson, T.R.; Carpenter, Z.L.; Hammons, D.R. Effects of Certain Electrical Stimulation Parameters on Quality and Palatability of Beef. *J. Food Sci.* **1981**, *46*, 13–18. [CrossRef]
5. Roeber, D.L.; Cannell, R.C.; Belk, K.E.; Tatum, J.D.; Smith, G.C. Effects of a unique application of electrical stimulation on tenderness, color, and quality attributes of the beef longissimus muscle. *J. Anim. Sci.* **2000**, *78*, 1504–1509. [CrossRef]
6. Li, C.; Li, J.; Li, X.; Hviid, M.; Lundström, K. Effect of low-voltage electrical stimulation after dressing on color stability and water holding capacity of bovine longissimus muscle. *Meat Sci.* **2011**, *88*, 559–565. [CrossRef]
7. Adeyemi, K.D.; Sazili, A.Q. Efficacy of Carcass Electrical Stimulation in Meat Quality Enhancement: A Review. *Asian-Australas. J. Anim. Sci.* **2014**, *27*, 447–456. [CrossRef] [PubMed]
8. McKeith, F.K.; Ellis, M.; Miller, K.D.; Sutton, D.S. The effect of RN genotype on pork quality. In Proceedings of the Reciprocal Meat Conference, Storrs, CT, USA, 12 May 1998; pp. 118–124.
9. Bowker, B.; Wynveen, E.; Grant, A.; Gerrard, D. Effects of electrical stimulation on early postmortem muscle pH and temperature declines in pigs from different genetic lines and halothane genotypes. *Meat Sci.* **1999**, *53*, 125–133. [CrossRef]
10. Wiklund, E.; Stevenson-Barry, J.; Duncan, S.; Littlejohn, R. Electrical stimulation of red deer (Cervus elaphus) carcasses—Effects on rate of pH-decline, meat tenderness, colour stability and water-holding capacity. *Meat Sci.* **2001**, *59*, 211–220. [CrossRef]

11. Kim, Y.; Lonergan, S.; Grubbs, J.; Cruzen, S.; Fritchen, A.; Della Malva, A.; Marino, R.; Huff-Lonergan, E. Effect of low voltage electrical stimulation on protein and quality changes in bovine muscles during postmortem aging. *Meat Sci.* **2013**, *94*, 289–296. [CrossRef] [PubMed]
12. McKenna, D.; Maddock, D.; Savell, J. Water-holding capacity and color characteristics of beef from electrically stimulated carcasses. *J. Muscle Foods* **2003**, *14*, 33–49. [CrossRef]
13. Nichols, J.E.; Cross, H.R. Effects of Electrical Stimulation and Early Postmortem Muscle Excision on pH Decline, Sarcomere Length and Color in Beef Muscles. *J. Food Prot.* **1980**, *43*, 514–519. [CrossRef]
14. Hwang, I.; Devine, C.; Hopkins, D. The biochemical and physical effects of electrical stimulation on beef and sheep meat tenderness. *Meat Sci.* **2003**, *65*, 677–691. [CrossRef]
15. Simmons, N.; Daly, C.; Cummings, T.; Morgan, S.; Johnson, N.; Lombard, A. Reassessing the principles of electrical stimulation. *Meat Sci.* **2008**, *80*, 110–122. [CrossRef] [PubMed]
16. Ding, C.; Rodas-González, A.; López-Campos, Ó.; Galbraith, J.; Juárez, M.; Larsen, I.; Jin, Y.; Aalhus, J. Effects of electrical stimulation on meat quality of bison striploin steaks and ground patties. *Can. J. Anim. Sci.* **2016**, *96*, 79–89. [CrossRef]
17. Dutson, T.; Smith, G.; Savell, J.; Carpenter, Z. Possible mechanisms by which electrical stimulation improves meat tenderness. In Proceedings of the 26th European Meeting of Meat Research Workers, Brno, Czechoslovakia, 12 May 1980; pp. 84–87.
18. Savell, J.; Dutson, T.; Smith, G.; Carpenter, Z. Structural Changes in Electrically Stimulated Beef Muscle. *J. Food Sci.* **1978**, *43*, 1606–1607. [CrossRef]
19. Ho, C.Y.; Stromer, M.H.; Robson, R.M. Effect of electrical stimulation on postmortem titin, nebulin, desmin, and troponin-T degradation and ultrastructural changes in bovine longissimus muscle. *J. Anim. Sci.* **1996**, *74*, 1563–1575. [CrossRef] [PubMed]
20. Juárez, M.; Basarab, J.A.; Baron, V.S.; Valera, M.; Óscar, L.-C.; Larsen, I.L.; Aalhus, J.L. Relative contribution of electrical stimulation to beef tenderness compared to other production factors. *Can. J. Anim. Sci.* **2016**, *96*, 104–107. [CrossRef]
21. Hopkins, D.; Thompson, J. Inhibition of protease activity 2. Degradation of myofibrillar proteins, myofibril examination and determination of free calcium levels. *Meat Sci.* **2001**, *59*, 199–209. [CrossRef]
22. Wheeler, T.L.; Shackelford, S.D.; Koohmaraie, M. Beef longissimus slice shear force measurement among steak locations and institutions1,2,3. *J. Anim. Sci.* **2007**, *85*, 2283–2289. [CrossRef]
23. Derington, A.; Brooks, J.; Garmyn, A.; Thompson, L.; Wester, D.; Miller, M. Relationships of slice shear force and Warner-Bratzler shear force of beef strip loin steaks as related to the tenderness gradient of the strip loin. *Meat Sci.* **2011**, *88*, 203–208. [CrossRef]
24. Shackelford, S.; Morgan, J.; Cross, H.; Savell, J. Identification of Threshold Levels for Warner-Bratzler Shear Force in Beef Top Loin Steaks. *J. Muscle Foods* **1991**, *2*, 289–296. [CrossRef]
25. Miller, M.F.; Hoover, L.C.; Cook, K.D.; Guerra, A.L.; Huffman, K.L.; Tinney, K.S.; Ramsey, C.B.; Brittin, H.C.; Huffman, L.M. Consumer Acceptability of Beef Steak Tenderness in the Home and Restaurant. *J. Food Sci.* **1995**, *60*, 963–965. [CrossRef]
26. Savell, J.; Smith, G.C.; Carpenter, Z.L. Beef quality and palatability as affected by electrical stimulation and cooler aging. *J. Food Sci.* **1978**, *43*, 1666–1668. [CrossRef]
27. Shanks, B.C.; Wulf, D.M.; Maddock, R.J. Technical note: The effect of freezing on Warner-Bratzler shear force values of beef longissimus steaks across several postmortem aging periods. *J. Anim. Sci.* **2002**, *80*, 2122–2125. [CrossRef] [PubMed]
28. McKeith, F.K.; Savell, J.W.; Smith, G.C. Tenderness Improvement of the Major Muscles of the Beef Carcass by Electrical Stimulation. *J. Food Sci.* **1981**, *46*, 1774–1776. [CrossRef]
29. Sleper, P.S.; Hunt, M.C.; Kropf, D.H.; Kastner, C.L.; Dikeman, M.E. Electrical Stimulation Effects on Myoglobin Properties of Bovine Longissimus Muscle. *J. Food Sci.* **1983**, *48*, 479–483. [CrossRef]
30. Zhang, Y.; Ji, X.; Mao, Y.; Luo, X.; Zhu, L.; Hopkins, D.L. Effect of new generation medium voltage electrical stimulation on the meat quality of beef slaughtered in a Chinese abattoir. *Meat Sci.* **2019**, *149*, 47–54. [CrossRef]

Article

Antimicrobial Polyamide-Alginate Casing Incorporated with Nisin and ε-Polylysine Nanoparticles Combined with Plant Extract for Inactivation of Selected Bacteria in Nitrite-Free Frankfurter-Type Sausage

Kazem Alirezalu [1,*], Milad Yaghoubi [2], Leila Poorsharif [2], Shadi Aminnia [3], Halil Ibrahim Kahve [4], Mirian Pateiro [5], José M. Lorenzo [5,6] and Paulo E. S. Munekata [5,*]

1 Department of Food Science and Technology, Ahar Faculty of Agriculture and Natural Resources, University of Tabriz, Tabriz 51666, Iran
2 Department of Food Science and Technology, Faculty of Agriculture, University of Tabriz, Tabriz 5166616471, Iran; m.yaghoubi97@ms.tabrizu.ac.ir (M.Y.); poorsharif.l@gmail.com (L.P.)
3 Department of Fisheries, Urmia University, Urmia 5756151818, Iran; sh.aminnia@urmia.ac.ir
4 Department of Food Engineering, Faculty of Engineering, Aksaray University, Aksaray 68100, Turkey; hibrahimkahve@gmail.com
5 Centro Tecnológico de la Carne de Galicia, Parque Tecnológico de Galicia, Avda. Galicia n 4, San Cibrao das Viñas, 32900 Ourense, Spain; mirianpateiro@ceteca.net (M.P.); jmlorenzo@ceteca.net (J.M.L.)
6 Área de Tecnología de los Alimentos, Facultad de Ciencias de Ourense, Universidad de Vigo, 32004 Ourense, Spain
* Correspondence: kazem.alirezalu@tabrizu.ac.ir (K.A.); paulosichetti@ceteca.net (P.E.S.M.)

Citation: Alirezalu, K.; Yaghoubi, M.; Poorsharif, L.; Aminnia, S.; Kahve, H.I.; Pateiro, M.; Lorenzo, J.M.; Munekata, P.E.S. Antimicrobial Polyamide-Alginate Casing Incorporated with Nisin and ε-Polylysine Nanoparticles Combined with Plant Extract for Inactivation of Selected Bacteria in Nitrite-Free Frankfurter-Type Sausage. Foods 2021, 10, 1003. https://doi.org/10.3390/foods10051003

Academic Editor: Salam A. Ibrahim

Received: 24 March 2021
Accepted: 1 May 2021
Published: 4 May 2021

Publisher's Note: MDPI stays neutral with regard to jurisdictional claims in published maps and institutional affiliations.

Copyright: © 2021 by the authors. Licensee MDPI, Basel, Switzerland. This article is an open access article distributed under the terms and conditions of the Creative Commons Attribution (CC BY) license (https://creativecommons.org/licenses/by/4.0/).

Abstract: The effects of combining a polyamide-alginate casing incorporated with nisin (100 ppm and 200 ppm) and ε-polylysine (500 ppm and 1000 ppm) nanoparticles and a mixed plant extract as ingredient in sausage formulation (500 ppm; composed of olive leaves (OLE), green tea (GTE) and stinging nettle extracts (SNE) in equal rates) were studied to improve the shelf life and safety of frankfurter-type sausage. The film characteristics and microbiological properties of sausage samples were evaluated. Sausage samples were packaged in polyethylene bags (vacuum condition) and analysed during 45 days of storage at 4 °C. Control sausages were also treated with 120 ppm sodium nitrite. Polyamide-alginate films containing 100 ppm nisin and 500 ε-PL nanoparticles had the highest ultimate tensile strength compared to other films. However, 100 ppm nisin and 500 ε-PL nanoparticles decreased water vapour permeability of films. The results also revealed that nisin nanoparticles had significantly ($p < 0.05$) low inhibitory effects against *Escherichia coli*, *Staphylococcus aureus*, molds and yeasts and total viable counts compared to control and ε-PL nanoparticles. Furthermore, 1000 ppm ε-PL nanoparticles displayed the highest antimicrobial activity. Based on the obtained results, the films containing ε-PL nanoparticle could be considered as a promising packaging for frankfurter-type sausages.

Keywords: polyamide-alginate film; nanoparticle; nisin; ε-polylysine; frankfurter; plant extract; active packaging

1. Introduction

Frankfurter-type sausages are a kind of emulsified meat products, which because of their high techno-functional components (bioavailable vitamins (B), essential amino acids, fatty acids, zinc, and heme-iron), ready to eat (RTE) product status, flavour acceptance and low cost, are widely consumed [1–3]. However, contamination in meat products like sausages, especially by foodborne bacteria, is the main concern of meat producers [4]. Moreover, frankfurter-type sausages with high fat content are sensitive to oxidation which leads to a reduction in quality (flavour and texture) during storage [5,6], hence researchers endeavour to reduce contamination in sausages using natural antimicrobials and antioxidants [7–9].

Increasing the safety of RTE products like sausages could be achieved using antimicrobial packaging [10–14]. Using natural antimicrobial compounds in packaging structures displays higher efficacy in comparison to direct use in the food matrix as an ingredient. The higher efficacy of active antimicrobial packaging may be attributed to the exposure of the food surface (where the risk of contamination is high) to antimicrobial compounds [15–17].

Polysaccharides like alginate due to their special colloidal properties such as film-forming, thickening, emulsion stabilizing agent, and gel producing are widely used as a biopolymer film or coating compounds [18]. For instance, sodium alginate has received great consideration due to promising properties in combination with calcium ions as delivery systems for active compounds [19–23]. In this regard, the U.S. Food and Drugs Administration (FDA) has indicated that alginate polymer has the Generally Recognized As Safe (GRAS) status for food use [24]. Furthermore, Surendhiran et al. [25] revealed that phlorotannin encapsulated in a alginate/poly(ethylene oxide) composite film inactivated *Salmonella* spp. and increased the shelf life of chicken meat.

Nisin is a nontoxic and stable bacteriocin produced from *Lactococcus lactis* with authorized use in food [26]. Nisin has strong antimicrobial properties against several spoilage bacteria in meat and meat products such as *Clostridia* and *Bacilli* spores [27,28]. In this regard, Churklam et al. [29] evaluated effects of carvacrol in combination with nisin on sliced Bologna sausage and they noticed that nisin and carvacrol inhibited microbial growth compared to control samples. Furthermore, other authors have reported similar results for nisin in ready-to-eat Yao meat products [30], pork loin [31], frankfurter-type sausage [6] and fresh sausage [32].

ε-Polylysine (ε-PL) is one of the most well-known natural components with high antimicrobial properties against a wide spectrum of bacteria like Gram-positive, Gram-negative bacteria (*Clostridium perfringens, Staphylococcus aureus* and *E. coli*), and yeast and molds [33]. ε-PL has high thermo-stability and is widely utilized in meat industry products like chilled beef [34], and frankfurter type sausage [35] as preservative. Furthermore, antimicrobial efficacy of ε-PL could be increased in combination with plant extracts [35]. Natural plant extracts like olive leaves (OLE), green tea (GTE) and stinging nettle (SNE) extracts are a good source of phenolic compounds with strong antimicrobial and antioxidant properties [36–38]. In OLE, the major polyphenols are oleuropein, oleuropein, hydroxytyrosol, tyrosol, luteolin-7-O-glucoside, apigenin-7-O-glucoside, p-coumaric acid, ferulic acid [39]. In the case of GTE the main polyphenols are catechin and its derivatives are epicatechin gallate, epigallocatechin gallate, and epigallocatechin [40]. The phenolic composition of SNE is comprised of flavonoids (kaempferol, rutin, isorhamnetin, and quercetin,) and phenolic acids (p-coumaric acid and ferulic acid) [41], making these plant materials rich sources of bioactive and nutrients compounds with potential ability as nitrite substitutes in frankfurter-type sausages [36].

The antimicrobial effect of combined ε-polylysine and nisin with natural antioixdants in frankfurter-type sausage has been reported [6]. To the best of our knowledge, there are no previous studies on the effect of active packaging containing ε-polylysine nanoparticles (ε-PLN) and nisin nanoparticles (NN) on the quality properties and stability of frankfurter-type sausage during storage. The aim of present study was thus to evaluate the effect of active polyamide-alginate films containing nisin and ε-PL nanoparticles in combination with plant extracts on quality and shelf life of frankfurter-type sausages.

2. Materials and Methods

2.1. Materials

All chemical components and microbial media were of analytical grade (purity >99%) and purchased from Merck (Darmstadt, Germany). Food-grade nisin (Nisaplin, 5000 IU/mL, and ε-polylysine powder (5000 IU/mL,) were purchased from Danisco (Copenhagen, Denmark) and FoodChemand (Shanghai, China), respectively. The high molecular mass chitosan powder (molecular weight 3.1×10^5 g mol^{-1}; 95% deacetylation degree) was also purchased from Sigma-Aldrich (Saint Louis, MO, USA). The polyamide film was acquired from Besharat Company (Tabriz, Iran).

2.2. Preparation of Plant Extract

The olive leaves (OLE), green tea (GTE) and stinging nettle (SNE) extracts were obtained following the procedure described by Ebrahimzadeh et al. [42] with some modifications. Plant leaves firstly dried in an oven (40 °C) for 48 h and sifted through sieves with 14-inch mesh. Then, 50 g of dried powder and 500 mL of ethanol solution (95%) in Erlenmeyer flask were mixed by magnetic stirrer at room temperature for 48 h. The ethanol in mixture was evaporated at 40 °C in a rotary evaporator after filtering the mixture through a Whatman No 1 filter paper. Equal amounts of GTE, SNE and OLE were prepared for further use in frankfurter-sausage samples.

2.3. Preparation of ε-Polylysine Nanoparticle (ε-PLN) and Nisin Nanoparticle (NN)

The stock solution of nisin and ε-PL were prepared according to method described by Alirezalu et al. [35] as follows: 2 g from each of these compounds were separately solubilized in 2% glacial acetic acid solution (200 mL) at 60 °C and filtered through 0.45 μm membrane filter (for sterilization) (Minisart NML, Sartorius, New York, NY, USA). The nanoparticles of nisin and ε-PL were produced according to the method described by Das et al. [43], and Bernela et al. [44] with some modifications. Calcium chloride and ε-PL solution (1:20, v/v) were mixed together. Then, this solution was mixed with 58.75 mL of sodium alginate (0.63 mg/mL), 12.5 mL of chitosan solution, and 6.25 mL of Pluronic F-68 (1 mg/mL). The final mixture was shaken slowly for 3.5 h (at room temperature). The nanoparticles of ε-PL were obtained after centrifugation (15,000× g) (at 4 °C for 0.5 h) and freeze-drying process. The similar technique was utilized for preparation of nisin nanoparticles.

2.4. Preparation of Active Antimicrobial Film

The polyamide-alginate films were produced according to the procedures described in [24,45]. Mechanical stirring was used to dissolve 3 g of sodium alginate in 200 mL of sterile deionised water for 30 min (70 °C). After that, for improving films characteristics (increase flexibility and decrease brittleness), glycerol (0.44 g/g of alginate) as a plasticizer was added. Mechanical stirring was utilized for dissolving glucono δ-lactone (5.4 g/g calcium carbonate) with calcium carbonate (0.03 g/g alginate) in 50 mL of distilled water. Then, for calcium alginate films production, the solution was dispersed at 150 mL sodium alginate solution. The ε-PLN (500 and 1000 ppm) and NN (100 and 200 ppm) were added to the mixture and homogenized for 3 min (13,500 rpm at 25 °C) and for integrate absorption of water and gelation of alginate, the attained solution was mixed slowly for 12 h by using mechanical stirring. After that, Petri dishes (10 cm in diagonal) were filled with 10 mL of the solution and dried in an oven (12 h at 45 °C). The Petri dishes were held in desiccators before peeling the films. Finally, polyamide and calcium alginate films were attached together to polyamide-alginate films.

2.5. Mechanical Properties

Mechanical properties of calcium alginate films including tensile strength (TS) and elongation at break (%E) were analysed following the ASTM procedure D882-91 [46]. The films were cut into 6 × 0.5 cm and were conditioned by saturated solution of calcium nitrite (RH = 55%) inside a desiccator for 24 h at room temperature. Initial grip separation with 50 mm and cross-head speed with 2 mm/min was used at this study with five replicates for mechanical analysis from each film samples.

2.6. Water Vapor Permeability (WVP)

The ASTM procedure E96-95 was utilized for gravimetrically evaluation of calcium alginate film WVP [46]. The calcium alginate films were sealed onto cups (2 cm diagonal and 10 cm height) with 3 g calcium sulfate and before hold inside desiccator containing saturated solution of potassium sulfate (RH = 97%) the cups were weighted. Finally, to attain a 97% RH gradient on the films the desiccator was placed inside the oven (25 °C). The cups were weighted twice a day (12 h to 12 h) for six days and the results were analysed by Fick and Henry's laws as follows:

$$WVP(\text{g mm/m2 day kPa}) = \frac{\text{The rate of water vapor transmission} \times \text{film thickness (mm)}}{\text{differential vapor pressure of water through the film}} \quad (1)$$

2.7. Preparation of Frankfurters

Each repetition of frankfurter processing was carried out with beef from different animals. The same ingredients and formulation were used in the three batches during three successive days (3 treatments × 4 time periods × 3 repetitions × 3 runs). A local meat processing factory was utilized for sausage production. Frankfurter-type sausage formulation (g/kg) was comprised of 0.4 sodium ascorbate, 15 salt, 81.5 starch and other dry materials, 120 soybean oil, 3.5 polyphosphate sodium, 20 seasoning, 210 ice/water, 0.5 mixed plant extract, and 550 of beef meat. Beef meat was cut into cubes with 3 mm size and homogenized with half of the ice/mixed extract (500 ppm), salt (NaCl) and sodium polyphosphate in a cutter (EX3000 RS, Kilia, Schönkirchen, Germany) at 10 °C for 12 min. After that, other ingredients including seasoning, starch, and sodium ascorbate were added slowly into the mix and homogenized for 1 min. Finally, half of ice/mixed extract (final mixed extract concentration of 500 ppm) and microbial suspensions (10^3 CFU/g), along with remaining components were added and mixed for about 120 s. The sausages were stuffed mechanically (VF50, Handtmann, Biberach, Germany) into antimicrobial polyamide-alginate films before steam cooking (1.5 h at 80–85 °C).

Sausages were quickly chilled with a cold-water shower, packaged in vacuumed condition in polyethylene bags and stored at 4 °C. The following five treatments were prepared: control with polyamide-alginate films without nanomaterials; samples stuffed in the polyamide-alginate films incorporated with 100 and 200 ppm NN, and samples stuffed in the polyamide-alginate films incorporated with 500 and 1000 ppm ε-PLN. Film characteristics and microbial counts were analysed at 0, 15, 30, and 45 days of refrigerated (4 ± 1 °C) storage. A schematic illustration to show the overall workflow is presented in Figure 1.

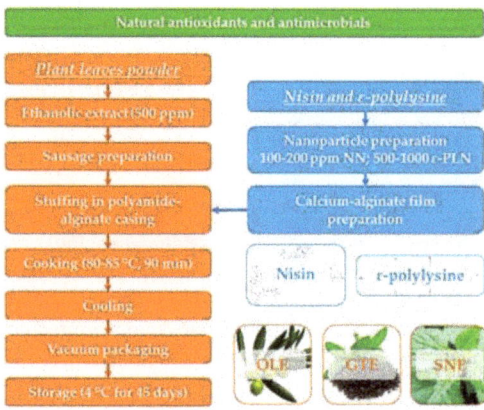

Figure 1. Schematic diagram for frankfurter-type sausage production.

2.8. Microbiological Properties

Microbiological properties of sausage packaged in polyamide alginate films were analysed as follows: 225 mL of 0.1% (w/v) peptone water and 25 g of samples were homogenized using sterile lab-blender (Paddle Lab Blender, Neutec, Farmingdale, NY, USA) for 3 min. 0.1% of sterile peptone water was also used for serial dilution production. Brilliant Green agar (BGA, Merck, Darmstadt, Germany), Plate Count agar (PCA, Merck) and Sulfite Polymyxin Sulfadizine (SPS) agar (Merck) were used for enumeration of *E. coli*, total viable count, and *Clostridium perfringens*, respectively, by pour-plate technique. Incubation time and temperature for *E. coli*, total viable count, and *Clostridium perfringens* were 24–47 h at 37 °C, 48–72 h at 30 °C and 24 h at 37 °C, respectively. *Staphylococcus aureus* and yeast and molds were enumerated on Baird Parker agar (BPA, Merck) and Dichloran Rose-Bengal Chloramphenicol (DRBC) agar (Merck) following incubation for 48 h at 30 °C and 5 days at 25 °C, respectively. The microbiological results were reported as Log10 CFU/g of sausage samples.

2.9. Statistical Analysis

The experimental data resulted from 3 treatments × 4 time periods × 3 repetitions × 3 runs were analysed using the statistical software SAS (v.9, SAS Institute Inc., Cary, NC, USA). Normal distribution and variance homogeneity had been previously tested (Shapiro-Wilk). Random block design was utilized for evaluation of microbiological data, considering a mixed linear model, including replication as a random effect, and different treatments and storage period as fixed impacts. One-way ANOVA was also utilized for mechanical properties and WVP, and Tukey's test for means comparison (statistical significance at $p < 0.05$ value) and results were expressed as mean values ± standard error in all figures and tables.

3. Results and Discussion

3.1. Mechanical Properties

The mechanical properties of calcium alginate films including elongation at break (E%) and tensile strength (TS) are shown at Table 1. The results showed that tensile strength of calcium alginate films ranged between 63–67 MPa. Added NN and ε-PLN affected significantly ($p < 0.05$) the tensile strength of films. Pranoto et al. [47] showed that tensile strength of calcium alginate films could decrease by adding garlic oil which may be caused by its hydrophobic properties. In our experiment, tensile strength in calcium alginate films incorporated with 500 ppm ε-PLN was higher than other films (Table 1). Conversely, ε-PLN (500 ppm ε-PLN and 1000 ppm ε-PLN) could significantly ($p < 0.05$) decrease the tensile strength of the films. This fact could be due to the high ε-PLN content, which can reduce tensile strength of the films. Benavides et al. [48] also reported similar results.

Table 1. Mechanical properties of active calcium alginate films.

	Treatments (ppm)					
Properties	Control	100 NN	200 NN	500 ε-PLN	1000 ε-PLN	*p*-Value
σ (MPa)	64.12 ± 2.38 [b]	66.59 ± 1.37 [a]	63.15 ± 0.81 [b]	67.08 ± 2.54 [a]	63.28 ± 1.78 [b]	0.03
E (%)	208.34 ± 2.34 [a]	206.54 ± 1.27 [ab]	207.65 ± 1.74 [ab]	205.47 ± 2.65 [b]	206.35 ± 2.73 [ab]	0.05

σ: Ultimate tensile strength (CV: 0.54); E: Elongation at break (CV: 1.71). [a-c] In each row with different letters differ significantly, ($p < 0.05$).

The results also showed that elongation at break (E%) in control films were higher than other films. Incorporating nanoparticles decreased the flexibility of the films, therefore, elongation at break were also decreased in the films. Moreover, link between alginate, calcium and chitosan decreased the flexibility and E% of the films. The results of this study are in agreement with those reported by Guiga et al. [49] in polyamide films. The authors indicated that added nisin in polyethylene and polyamide films could decrease E% of the films from 271% to 130%. The appearance of the active films used are presented in Figure 2.

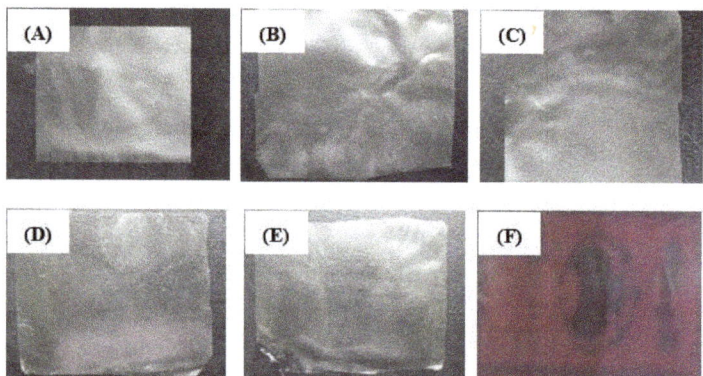

Figure 2. Calcium alginate films and polyamide-alginate film incorporated with NN and ε-PLN. (**A**) Calcium alginate film, (**B**) Calcium alginate film + 100 ppm NN, (**C**) Calcium alginate film + 200 ppm NN, (**D**) Calcium alginate film + 500 ppm ε-PLN, (**E**) Calcium alginate film + 1000 ppm ε-PLN, (**F**) Polyamide-alginate film.

3.2. Water Vapor Permeability (WVP)

The effects of nanoparticles in WVP of calcium alginate films are shown in Figure 3. According to Barzegaran et al. [50], calcium alginate films have lower water vapour permeability (6.16×10^{-7} g/m.s.Pa) in comparison to other polymers. Therefore, for production of films with antimicrobial activities with low water vapour permeability, calcium alginate films were produced. The results showed that adding 100 ppm of NN in calcium alginate films decreased WVP of the films, whereas incorporating 200 ppm of NN increased the WVP of the films.

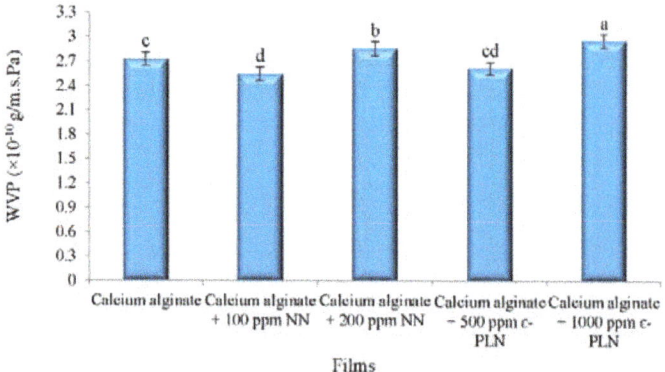

Figure 3. Water vapour permeability of calcium alginate films incorporated with nisin and ε-PL nanoparticles. a–d Mean values among films not followed by a common letter differ significantly ($p < 0.05$).

Sodium alginate films can be considered as an edible casing because of its hydrophilic properties (low resistant against moisture) and mechanical stability [51]. Therefore, favouring the gel consistency in films (by adding calcium) can improve water vapour permeability of the films [52].

WVP of the alginate films were decreased by adding 500 ppm ε-PLN. The interaction between amines (chitosan) and carboxyl components (alginate) may be the main reason for the lower WVP in films. Intermolecular gaps and porous microstructure of films matrix significantly affect the permeability of the films. Turhan and Şahbaz [53] showed that plasticizers can increase water vapour permeability of the films by increasing intermolecular gaps. As shown in Figure 3, WVP of films incorporated with 1000 ppm ε-PLN and 200 ppm NN were significantly increased whereas the WVP of films with 500 ppm ε-PLN and 100 ppm NN were decreased. These results may be caused by increasing and decreasing intermolecular gaps, respectively.

3.3. Microbiological Properties

Total viable count in all sausage samples ($p < 0.05$) increased during storage. The polyamide-alginate films incorporated with 1000 ppm ε-PLN and 200 ppm NN showed ($p < 0.05$) higher inhibitory effects against TVC. Furthermore, polyamide-alginate films incorporated with ε-PLN presented higher antimicrobial effects compared to polyamide-alginate films with NN. The antimicrobial effects of ε-PL against wide spectrum of microorganisms (Gram-positive, Gram-negative, and fungus) compared to nisin may be the main reason that explains the higher antimicrobial properties of ε-PLN.

TVC of frankfurter sausages packaged in polyamide-alginate films with 500 ppm ε-PLN and 100 ppm NN ranged between 5.54 and 5.87 Log CFU/g at the end of storage period (Figure 4). It is worth mentioning that the borderline for microbiological acceptability in meat products (especially due to odour changes) is around 6 Log CFU/g [54,55]. Conversely, TVC in sausage samples packaged in films with 1000 ppm ε-PLN and 200 ppm NN reached 4.22 Log CFU/g and 4.53 Log CFU/g, respectively. Therefore, the results showed that polyamide-alginate films incorporated with 500 and 1000 ppm ε-PLN could ($p < 0.05$) can increase the shelf life of the frankfurter-type sausages (Figure 4). In this regard, Feng et al. [56] indicated that ε-PL with rosemary extract could significantly decrease TVC and improve sensory properties of chicken breast muscle. Additionally, Alirezalu et al. [35] also indicated that ε-PLN displayed significantly higher inhibitory activity against TVC in comparison ε-PL (free form) in frankfurter-type sausage.

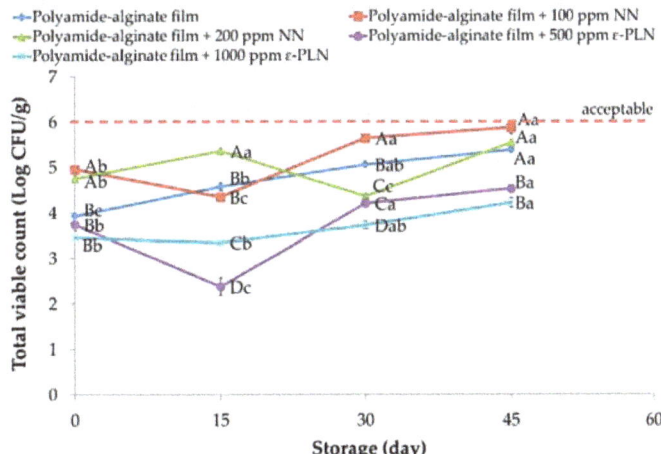

Figure 4. Total viable count of sausage samples packaged in polyamide-alginate films during refrigerated storage. [A–D] Mean values among treatments not followed by a common letter differ significantly ($p < 0.05$). [a–c] Mean values during storage not followed by a common letter differ significantly ($p < 0.05$).

Conversely, de Barros et al. [57] evaluated effects of natural casing incorporated with nisin in vacuum packaged sausage for the control of spoilage microorganisms, and reported an inhibitory effects of nisin against TVC. Similar results supporting the antimicrobial activity of nisin were also reported by Neetoo and Mahomoodally [58] on cold smoked salmon (by using cellulose-based films and coatings incorporated with nisin and potassium sorbate), and Ercolini et al. [59] on beef burgers coated in nisin and packaged in LDPE films.

Regarding the antimicrobial effect of films against *Clostridium perfringens*, counts between 2.43 and 2.86 Log CFU/g in all sausage samples were obtained at the first day of storage (Figure 5). During storage, significant reductions were observed among treatments and at the of storage the treatments control, polyamide-alginate films with 500 and 1000 ppm ε-PLN had lower values than those sausages packaged with polyamide-algine films with 100 and 200 NN. However, significant differences between sausages packaged in polyamide-alginate films containing 1000 ppm ε-PLN and control group after 45 days of refrigerated storage.

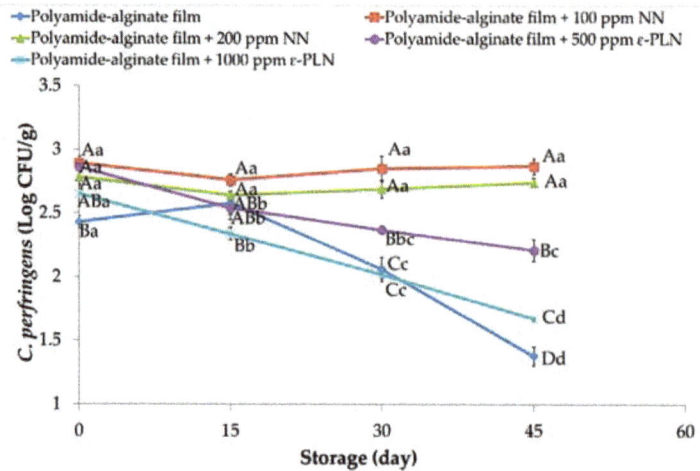

Figure 5. *Clostridium perfringens* of sausage samples packaged in polyamide-alginate films during refrigerated storage. $^{A-D}$ Mean values among treatments not followed by a common letter differ significantly ($p < 0.05$). $^{a-c}$ Mean values during storage not followed by a common letter differ significantly ($p < 0.05$).

Meira et al. [60] evaluated antimicrobial effects of polypropylene/montmorillonite nanocomposites containing different concentration of nisin as antimicrobial active packaging. The authors showed that nisin inhibited the growth of *Clostridium perfringens*, which was stronger in samples with higher concentrations of nisin. In addition, Cé et al. [61] also reported similar results against *Clostridium perfringens* in chitosan films containing nisin. It is important to comment that *Clostridium perfringens* is an anaerobic bacterium that mostly growth in inner sections of the sausage and the antimicrobial were in films that are in contact with the external surface of the samples. Therefore, our results did not show high inhibitory effects against *Clostridium perfringens* in sausage samples.

Staphylococcus aureus in meat and meat products is one of the most important bacteria because of its enterotoxin production [28]. Polyamide-alginate films incorporated with NN and ε-PLN had significant ($p < 0.05$) effects on *Staphylococcus aureus* count (Figure 6). During the refrigerated storage, *Staphylococcus aureus* significantly ($p < 0.05$) decreased in all sausage samples. Our results revealed that *Staphylococcus aureus* counts in samples packaged in polyamide-alginate films incorporated with 500 ppm and 1000 ppm ε-PLN and control sausages were significantly ($p < 0.05$) lower than obtained in other sausages. Elmani [62] evaluated the antimicrobial effects of lysozyme, chitosan, and nisin on Cig kofte (a traditional Turkish raw meatball) and showed that the inhibitory effect of nisin against *Staphylococcus aureus* was higher than chitosan and lysozyme. Our findings agree with data reported by Millette et al. [63] who reported alginate films containing 1000 IU/mL of nisin could decrease 2 Log CFU/cm² of *Staphylococcus aureus* in beef meat after 7 days of storage.

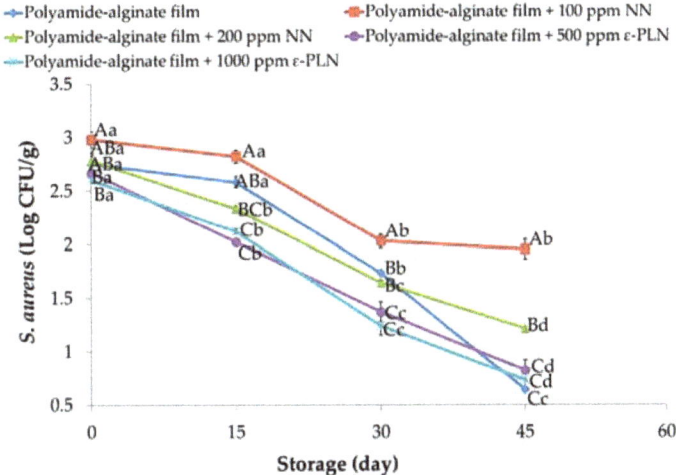

Figure 6. *Staphylococcus aureus* of sausage samples packaged in polyamide-alginate films during refrigerated storage. $^{A-C}$ Mean values among treatments not followed by a common letter differ significantly ($p < 0.05$). $^{a-d}$ Mean values during storage not followed by a common letter differ significantly ($p < 0.05$).

ε-PLN significantly ($p < 0.05$) affected *E. coli* in sausage samples (Figure 7). *E. coli* counts continuously decreased in all packaged sausages during storage except for samples packaged with NN ($p > 0.05$) due to lower antimicrobial effects of nisin against Gram-negative bacteria.

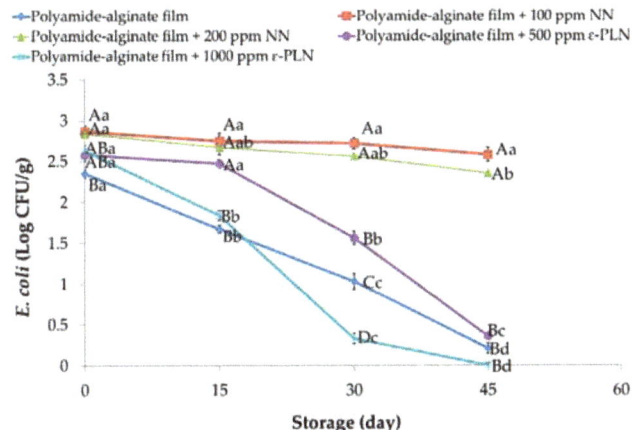

Figure 7. *E. coli* of sausage samples packaged in polyamide-alginate films during refrigerated storage. $^{A-D}$ Mean values among treatments not followed by a common letter differ significantly ($p < 0.05$). $^{a-d}$ Mean values during storage not followed by a common letter differ significantly ($p < 0.05$).

The presence of chitosan in nisin nanoparticles' structure leads to low inhibitory effects of polyamide-alginate films containing nisin against *E. coli*. Our outcomes agree with those reported by Cé et al. [61] who observed that higher concentrations of chitosan had weaker inhibitory effects against *E. coli*. Furthermore, Elmali [62] evaluated the antimicrobial effects of lysozyme, chitosan, and nisin on Cig kofte and reported there were any inhibitory effects against *E. coli* in samples treated with nisin after 72 h of storage. Polyamide-alginate films containing ε-PLN decreased ($p < 0.05$) *E. coli* in sausage samples during the storage time.

Our results showed that *E. coli* count in sausages packaged in films containing 1000 ppm ε-PLN reached 0 Log CFU/g after 45 days of storage. The high antimicrobial effects of ε-PLN against *E. coli* were also reported by Sun et al. [64] who evaluated the antimicrobial effects of nano-crystalline cellulose films containing ε-PL on fish meat.

At day 1, sausage samples packaged in polyamide-alginate films with NN and ε-PLN presented ($p < 0.05$) lower content of molds and yeasts in comparison to control group (Figure 8). During storage, molds and yeasts increased in all sausage samples. At the end of storage time, sausage samples packaged in films containing 100 ppm NN showed the highest counts of molds and yeasts. Our findings agree with data reported by Guerra et al. [65] who evaluated the antimicrobial effects of cellophane containing nisin and reported that bioactive cellophane packaging could be used for controlling microbial growth in chopped meat. Packaged sausage samples in polyamide-alginate containing ε-PLN (500 and 1000 ppm) decreased ($p < 0.05$) the rate on molds and yeasts growth during refrigerated storage. Furthermore, in control samples (with 120 ppm sodium nitrite) molds and yeasts counts until day 30 were within the standard range.

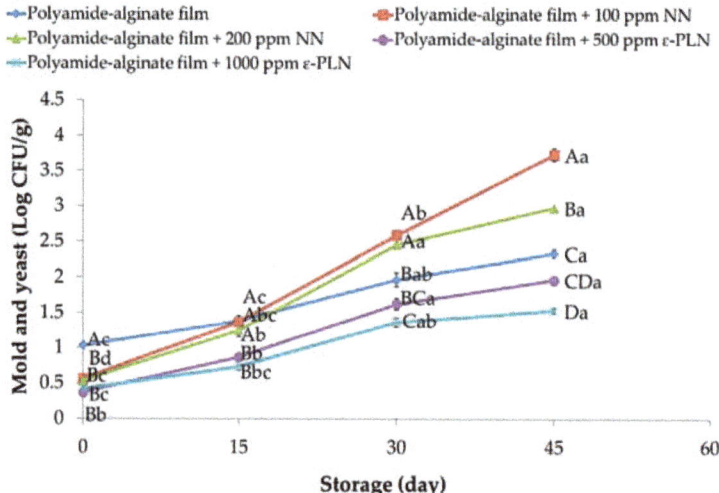

Figure 8. Molds and yeasts of sausage samples packaged in polyamide-alginate films during refrigerated storage. [A–D] Mean values among treatments not followed by a common letter differ significantly ($p < 0.05$). [a–d] Mean values during storage not followed by a common letter differ significantly ($p < 0.05$).

Alirezalu et al. [6] also reported similar results in frankfurter sausages. These authors evaluated the antimicrobial effects of nisin, ε-PL and chitosan and reported a similar inhibitory effect against molds and yeasts in meat products. Moreover, catechins in GTE as a phenolic compound not only can inhibit activity of intracellular enzymes and synthesis of fatty acid and protein but also can damage membrane compounds of molds and yeasts [66].

4. Conclusions

Our outcomes showed that polyamide-alginate casing incorporated with ε-PLN and NN with mixed plant extract (same rates of green tea, stinging nettle, and olive leaves extracts) could be potentially used for increasing frankfurter-type sausage shelf life. Sausages with 1000 ppm ε-PLN had significantly higher inhibitory effects against molds and yeasts,

E. coli, *Staphylococcus aureus*, and total viable counts. Therefore, polyamide-alginate film incorporated with ε-PLN and NN with mixed plant extract could be used usefully for improving frankfurter type sausage quality and shelf life.

Author Contributions: Conceptualization, K.A., M.Y. and H.I.K.; writing—original draft preparation, K.A. and L.P.; writing—review and editing, K.A., M.Y., S.A., H.I.K., M.P., J.M.L., and P.E.S.M. All authors have read and agreed to the published version of the manuscript.

Funding: This research received no external funding.

Data Availability Statement: The data presented in this study are available on request from the corresponding author.

Acknowledgments: Thanks to GAIN (Axencia Galega de Innovación) for supporting this review (grant number IN607A2019/01). José M. Lorenzo, Paulo E. S. Munekata and Mirian Pateiro are members of the HealthyMeat network, funded by CYTED (ref. 119RT0568).

Conflicts of Interest: The authors declare no conflict of interest.

References

1. Novakovic, S.; Djekic, I.; Klaus, A.; Vunduk, J.; Djordjevic, V.; Tomović, V.; Šojić, B.; Kocić-Tanackov, S.; Lorenzo, J.M.; Barba, F.J.; et al. The effect of *Cantharellus cibarius* addition on quality characteristics of frankfurter during refrigerated storage. *Foods* **2019**, *8*, 635. [CrossRef]
2. Franco, D.; Martins, A.J.; López-Pedrouso, M.; Purriños, L.; Cerqueira, M.A.; Vicente, A.A.; Pastrana, L.M.; Zapata, C.; Lorenzo, J.M. Strategy towards replacing pork backfat with a linseed oleogel in frankfurter sausages and its evaluation on physicochemical, nutritional, and sensory characteristics. *Foods* **2019**, *8*, 366. [CrossRef] [PubMed]
3. Henck, J.M.M.; Bis-Souza, C.V.; Pollonio, M.A.R.; Lorenzo, J.M.; Barretto, A.C.S. Alpha-cyclodextrin as a new functional ingredient in low-fat chicken frankfurter. *Br. Poult. Sci.* **2019**, *60*, 716–723. [CrossRef] [PubMed]
4. Fougy, L.; Desmonts, M.H.; Coeuret, G.; Fassel, C.; Hamon, E.; Hézard, B.; Champomier-Vergès, M.C.; Chaillou, S. Reducing salt in raw pork sausages increases spoilage and correlates with reduced bacterial diversity. *Appl. Environ. Microbiol.* **2016**, *82*, 3928–3939. [CrossRef] [PubMed]
5. Domínguez, R.; Pateiro, M.; Agregán, R.; Lorenzo, J.M. Effect of the partial replacement of pork backfat by microencapsulated fish oil or mixed fish and olive oil on the quality of frankfurter type sausage. *J. Food Sci. Technol.* **2017**, *54*, 26–37. [CrossRef] [PubMed]
6. Alirezalu, K.; Hesari, J.; Nemati, Z.; Munekata, P.E.S.; Barba, F.J.; Lorenzo, J.M. Combined effect of natural antioxidants and antimicrobial compounds during refrigerated storage of nitrite-free frankfurter-type sausage. *Food Res. Int.* **2019**, *120*, 839–850. [CrossRef]
7. Lorenzo, J.M.; González-Rodríguez, R.M.; Sánchez, M.; Amado, I.R.; Franco, D. Effects of natural (grape seed and chestnut extract) and synthetic antioxidants (buthylatedhydroxytoluene, BHT) on the physical, chemical, microbiological and sensory characteristics of dry cured sausage "chorizo". *Food Res. Int.* **2013**, *54*, 611–620. [CrossRef]
8. de Carvalho, F.A.L.; Munekata, P.E.S.; Lopes de Oliveira, A.; Pateiro, M.; Domínguez, R.; Trindade, M.A.; Lorenzo, J.M. Turmeric (*Curcuma longa* L.) extract on oxidative stability, physicochemical and sensory properties of fresh lamb sausage with fat replacement by tiger nut (*Cyperus esculentus* L.) oil. *Food Res. Int.* **2020**, *136*, 109487. [CrossRef]
9. Fernandes, R.P.P.; Trindade, M.A.; Lorenzo, J.M.; de Melo, M.P. Assessment of the stability of sheep sausages with the addition of different concentrations of *Origanum vulgare* extract during storage. *Meat Sci.* **2018**, *137*, 244–257. [CrossRef]
10. Gómez, M.; Lorenzo, J.M. Effect of packaging conditions on shelf-life of fresh foal meat. *Meat Sci.* **2012**, *91*, 513–520. [CrossRef]
11. Domínguez, R.; Barba, F.J.; Gómez, B.; Putnik, P.; Bursać Kovačević, D.; Pateiro, M.; Santos, E.M.; Lorenzo, J.M. Active packaging films with natural antioxidants to be used in meat industry: A review. *Food Res. Int.* **2018**, *113*, 93–101. [CrossRef]
12. Umaraw, P.; Munekata, P.E.S.; Verma, A.K.; Barba, F.J.; Singh, V.P.; Kumar, P.; Lorenzo, J.M. Edible films/coating with tailored properties for active packaging of meat, fish and derived products. *Trends Food Sci. Technol.* **2020**, *98*, 10–24. [CrossRef]
13. Horita, C.N.; Baptista, R.C.; Caturla, M.Y.R.; Lorenzo, J.M.; Barba, F.J.; Sant'Ana, A.S. Combining reformulation, active packaging and non-thermal post-packaging decontamination technologies to increase the microbiological quality and safety of cooked ready-to-eat meat products. *Trends Food Sci. Technol.* **2018**, *72*, 45–61. [CrossRef]
14. Pateiro, M.; Domínguez, R.; Bermúdez, R.; Munekata, P.E.S.; Zhang, W.; Gagaoua, M.; Lorenzo, J.M. Antioxidant active packaging systems to extend the shelf life of sliced cooked ham. *Curr. Res. Food Sci.* **2019**, *1*, 24–30. [CrossRef] [PubMed]
15. Pateiro, M.; Munekata, P.E.S.; Sant'Ana, A.S.; Domínguez, R.; Rodríguez-Lázaro, D.; Lorenzo, J.M. Application of essential oils as antimicrobial agents against spoilage and pathogenic microorganisms in meat products. *Int. J. Food Microbiol.* **2021**, *337*, 108966. [CrossRef]
16. Pateiro, M.; Vargas, F.C.; Chincha, A.A.I.A.; Sant'Ana, A.S.; Strozzi, I.; Rocchetti, G.; Barba, F.J.; Domínguez, R.; Lucini, L.; do Amaral Sobral, P.J.; et al. Guarana seed extracts as a useful strategy to extend the shelf life of pork patties: UHPLC-ESI/QTOF phenolic profile and impact on microbial inactivation, lipid and protein oxidation and antioxidant capacity. *Food Res. Int.* **2018**, *114*, 55–63. [CrossRef]

17. Nikmaram, N.; Budaraju, S.; Barba, F.J.; Lorenzo, J.M.; Cox, R.B.; Mallikarjunan, K.; Roohinejad, S. Application of plant extracts to improve the shelf-life, nutritional and health-related properties of ready-to-eat meat products. *Meat Sci.* **2018**, *145*, 245–255. [CrossRef]
18. Mohamed, S.A.A.; El-Sakhawy, M.; El-Sakhawy, M.A.M. Polysaccharides, protein and lipid-based natural edible films in food packaging: A review. *Carbohydr. Polym.* **2020**, *238*, 116178. [CrossRef] [PubMed]
19. Acevedo-Fani, A.; Salvia-Trujillo, L.; Rojas-Graü, M.A.; Martín-Belloso, O. Edible films from essential-oil-loaded nanoemulsions: Physicochemical characterization and antimicrobial properties. *Food Hydrocoll.* **2015**, *47*, 168–177. [CrossRef]
20. Rezaei, F.; Shahbazi, Y. Shelf-life extension and quality attributes of sauced silver carp fillet: A comparison among direct addition, edible coating and biodegradable film. *LWT Food Sci. Technol.* **2018**, *87*, 122–133. [CrossRef]
21. Kristam, P.; Eswarapragada, N.M.; Bandi, E.R.; Tumati, S.R. Evaluation of edible polymer coatings enriched with green tea extract on quality of chicken nuggets. *Vet. World* **2016**, *9*, 685–692. [CrossRef] [PubMed]
22. Aluani, D.; Tzankova, V.; Kondeva-Burdina, M.; Yordanov, Y.; Nikolova, E.; Odzhakov, F.; Apostolov, A.; Markova, T.; Yoncheva, K. Evaluation of biocompatibility and antioxidant efficiency of chitosan-alginate nanoparticles loaded with quercetin. *Int. J. Biol. Macromol.* **2017**, *103*, 771–782. [CrossRef] [PubMed]
23. Cai, L.; Cao, A.; Bai, F.; Li, J. Effect of ε-polylysine in combination with alginate coating treatment on physicochemical and microbial characteristics of Japanese sea bass (Lateolabrax japonicas) during refrigerated storage. *LWT Food Sci. Technol.* **2015**, *62*, 1053–1059. [CrossRef]
24. Rhim, J.W. Physical and mechanical properties of water resistant sodium alginate films. *LWT Food Sci. Technol.* **2004**, *37*, 323–330. [CrossRef]
25. Surendhiran, D.; Cui, H.; Lin, L. Encapsulation of Phlorotannin in Alginate/PEO blended nanofibers to preserve chicken meat from Salmonella contaminations. *Food Packag. Shelf Life* **2019**, *21*, 100346. [CrossRef]
26. Correa, J.P.; Molina, V.; Sanchez, M.; Kainz, C.; Eisenberg, P.; Massani, M.B. Improving ham shelf life with a polyhydroxybutyrate/polycaprolactone biodegradable film activated with nisin. *Food Packag. Shelf Life* **2017**, *11*, 31–39. [CrossRef]
27. Krivorotova, T.; Cirkovas, A.; Maciulyte, S.; Staneviciene, R.; Budriene, S.; Serviene, E.; Sereikaite, J. Nisin-loaded pectin nanoparticles for food preservation. *Food Hydrocoll.* **2016**, *54*, 49–56. [CrossRef]
28. Lorenzo, J.M.; Munekata, P.E.S.; Dominguez, R.; Pateiro, M.; Saraiva, J.A.; Franco, D. Main Groups of Microorganisms of Relevance for Food Safety and Stability: General Aspects and Overall Description. In *Innovative Technologies for Food Preservation Inactivation of Spoilage and Pathogenic Microorganisms*; Barba, F.J., Sant'Ana, A.S., Orlie, V., Koubaa, M., Eds.; Academic Press: London, UK, 2017; pp. 53–107, ISBN 9780128110324.
29. Churklam, W.; Chaturongakul, S.; Ngamwongsatit, B.; Aunpad, R. The mechanisms of action of carvacrol and its synergism with nisin against Listeria monocytogenes on sliced bologna sausage. *Food Control* **2020**, *108*, 106864. [CrossRef]
30. Liu, Q.; Zhang, M.; Bhandari, B.; Xu, J.; Yang, C. Effects of nanoemulsion-based active coatings with composite mixture of star anise essential oil, polylysine, and nisin on the quality and shelf life of ready-to-eat Yao meat products. *Food Control* **2020**, *107*, 106771. [CrossRef]
31. Cao, Y.; Warner, R.D.; Fang, Z. Effect of chitosan/nisin/gallic acid coating on preservation of pork loin in high oxygen modified atmosphere packaging. *Food Control* **2019**, *101*, 9–16. [CrossRef]
32. Araújo, M.K.; Gumiela, A.M.; Bordin, K.; Luciano, F.B.; de Macedo, R.E.F. Combination of garlic essential oil, allyl isothiocyanate, and nisin Z as bio-preservatives in fresh sausage. *Meat Sci.* **2018**, *143*, 177–183. [CrossRef]
33. Pandey, A.K.; Kumar, A. Improved microbial biosynthesis strategies and multifarious applications of the natural biopolymer epsilon-poly-l-lysine. *Process Biochem.* **2014**, *49*, 496–505. [CrossRef]
34. Alirezalu, K.; Movlan, H.S.; Yaghoubi, M.; Pateiro, M.; Lorenzo, J.M. ε-polylysine coating with stinging nettle extract for fresh beef preservation. *Meat Sci.* **2021**, *176*, 108474. [CrossRef] [PubMed]
35. Alirezalu, K.; Hesari, J.; Yaghoubi, M.; Khaneghah, A.M.; Alirezalu, A.; Pateiro, M.; Lorenzo, J.M. Combined effects of ε-polylysine and ε-polylysine nanoparticles with plant extracts on the shelf life and quality characteristics of nitrite-free frankfurter-type sausages. *Meat Sci.* **2021**, *172*, 108318. [CrossRef] [PubMed]
36. Alirezalu, K.; Pateiro, M.; Yaghoubi, M.; Alirezalu, A.; Peighambardoust, S.H.; Lorenzo, J.M. Phytochemical constituents, advanced extraction technologies and techno-functional properties of selected Mediterranean plants for use in meat products. A comprehensive review. *Trends Food Sci. Technol.* **2020**, *100*, 292–306. [CrossRef]
37. Lorenzo, J.M.; Munekata, P.E.S. Phenolic compounds of green tea: Health benefits and technological application in food. *Asian Pac. J. Trop. Biomed.* **2016**, *6*, 709–719. [CrossRef]
38. Sahin, S.; Samli, R.; Birteks Z Tan, A.S.; Barba, F.J.; Chemat, F.; Cravotto, G.; Lorenzo, J.M. Solvent-free microwave-assisted extraction of polyphenols from olive tree leaves: Antioxidant and antimicrobial properties. *Molecules* **2017**, *22*, 1056. [CrossRef] [PubMed]
39. Hayes, J.E.; Allen, P.; Brunton, N.; O'Grady, M.N.; Kerry, J.P. Phenolic composition and in vitro antioxidant capacity of four commercial phytochemical products: Olive leaf extract (Olea europaea L.), lutein, sesamol and ellagic acid. *Food Chem.* **2011**, *126*, 948–955. [CrossRef]
40. Zhao, C.N.; Tang, G.Y.; Cao, S.Y.; Xu, X.Y.; Gan, R.Y.; Liu, Q.; Mao, Q.Q.; Shang, A.; Li, H. Bin Phenolic profiles and antioxidant activities of 30 tea infusions from green, black, oolong, white, yellow and dark teas. *Antioxidants* **2019**, *8*, 215. [CrossRef]

41. Komes, D.; Belščak-Cvitanović, A.; Horžić, D.; Rusak, G.; Likić, S.; Berendika, M. Phenolic composition and antioxidant properties of some traditionally used medicinal plants affected by the extraction time and hydrolysis. *Phytochem. Anal.* **2011**, *22*, 172–180. [CrossRef]
42. Ebrahimzadeh, M.A.; Pourmorad, F.; Hafezi, S. Antioxidant activities of Iranian corn silk. *Turkish J. Biol.* **2008**, *32*, 43–49.
43. Das, R.K.; Kasoju, N.; Bora, U. Encapsulation of curcumin in alginate-chitosan-pluronic composite nanoparticles for delivery to cancer cells. *Nanomed. Nanotechnol. Biol. Med.* **2010**, *6*, 153–160. [CrossRef] [PubMed]
44. Bernela, M.; Kaur, P.; Chopra, M.; Thakur, R. Synthesis, characterization of nisin loaded alginate-chitosan-pluronic composite nanoparticles and evaluation against microbes. *LWT Food Sci. Technol.* **2014**, *59*, 1093–1099. [CrossRef]
45. Ingar Draget, K.; Østgaard, K.; Smidsrød, O. Homogeneous alginate gels: A technical approach. *Carbohydr. Polym.* **1990**, *14*, 159–178. [CrossRef]
46. ASTM ASTM E96/E96M-16, standard test methods for water vapor transmission of materials. In *Annual Book of American Standard Testing Methods*; American Society for Testing and Materials: West Conshohocken, PA, USA, 2016; pp. 719–725.
47. Pranoto, Y.; Salokhe, V.M.; Rakshit, S.K. Physical and antibacterial properties of alginate-based edible film incorporated with garlic oil. *Food Res. Int.* **2005**, *38*, 267–272. [CrossRef]
48. Benavides, S.; Villalobos-Carvajal, R.; Reyes, J.E. Physical, mechanical and antibacterial properties of alginate film: Effect of the crosslinking degree and oregano essential oil concentration. *J. Food Eng.* **2012**, *110*, 232–239. [CrossRef]
49. Guiga, W.; Galland, S.; Peyrol, E.; Degraeve, P.; Carnet-Pantiez, A.; Sebti, I. Antimicrobial plastic film: Physico-chemical characterization and nisin desorption modeling. *Innov. Food Sci. Emerg. Technol.* **2009**, *10*, 203–207. [CrossRef]
50. Barzegaran, A.; Jokar, M.; Dakheli, M.J. Effects of Green Tea Extract on Physicochemical and Antioxidant Properties of Polyamide Packaging Film. *J. Chem. Heal. Risks* **2014**, *4*, 41–48. [CrossRef]
51. Han, J.H.; Aristippos, G. Edible films and coatings. A review. In *Innovations in Food Packaging*; Han, J.H., Ed.; Elsevier Ltd.: San Diego, CA, USA, 2005; pp. 239–262. ISBN 9780123116321.
52. Pavlath, A.E.; Gossett, C.; Camirand, W.; Robertson, G.H. Ionomeric Films of Alginic Acid. *J. Food Sci.* **1999**, *64*, 61–63. [CrossRef]
53. Turhan, K.N.; Şahbaz, F. Water vapor permeability, tensile properties and solubility of methylcellulose-based edible films. *J. Food Eng.* **2004**, *61*, 459–466. [CrossRef]
54. Food Safety Authority of Ireland. Available online: https://www.fsai.ie/food_businesses/micro_criteria/guideline_micro_criteria.html (accessed on 5 March 2021).
55. Dainty, R.H.; Mackey, B.M. The relationship between the phenotypic properties of bacteria from chill-stored meat and spoilage processes. *J. Appl. Bacteriol.* **1992**, *73*, 103s–114s. [CrossRef] [PubMed]
56. Feng, L.; Shi, C.; Bei, Z.; Li, Y.; Yuan, D.; Gong, Y.; Han, J. Rosemary Extract in Combination with ε-Polylysine Enhance the Quality of Chicken Breast Muscle during Refrigerated Storage. *Int. J. Food Prop.* **2016**, *19*, 2338–2348. [CrossRef]
57. de Barros, J.R.; Kunigk, L.; Jurkiewicz, C.H. Incorporation of nisin in natural casing for the control of spoilage microorganisms in vacuum packaged sausage. *Braz. J. Microbiol.* **2010**, *41*, 1001–1008. [CrossRef]
58. Neetoo, H.; Mahomoodally, F. Use of antimicrobial films and edible coatings incorporating chemical and biological preservatives to control growth of Listeria monocytogenes on cold smoked salmon. *Biomed Res. Int.* **2014**, *2014*. [CrossRef]
59. Ercolini, D.; Ferrocino, I.; La Storia, A.; Mauriello, G.; Gigli, S.; Masi, P.; Villani, F. Development of spoilage microbiota in beef stored in nisin activated packaging. *Food Microbiol.* **2010**, *27*, 137–143. [CrossRef] [PubMed]
60. Meira, S.M.M.; Zehetmeyer, G.; Jardim, A.I.; Scheibel, J.M.; de Oliveira, R.V.B.; Brandelli, A. Polypropylene/montmorillonite nanocomposites containing nisin as antimicrobial food packaging. *Food Bioprocess Technol.* **2014**, *7*, 3349–3357. [CrossRef]
61. Cé, N.; Noreña, C.P.Z.; Brandelli, A. Antimicrobial activity of chitosan films containing nisin, peptide P34, and natamycin. *CyTA J. Food* **2012**, *10*, 21–26. [CrossRef]
62. Elmalı, M. Effects of different concentration of nisin, lysozyme, and chitosan on the changes of microorganism profile in produced Çiğ Köfte (Turkish traditional meat product; raw meatball) during the production stage. *MANAS J. Eng.* **2014**, *2*, 30–45.
63. Millette, M.; Le Tien, C.; Smoragiewicz, W.; Lacroix, M. Inhibition of Staphylococcus aureus on beef by nisin-containing modified alginate films and beads. *Food Control* **2007**, *18*, 878–884. [CrossRef]
64. Sun, X.; Guo, X.; Ji, M.; Wu, J.; Zhu, W.; Wang, J.; Cheng, C.; Chen, L.; Zhang, Q. Preservative effects of fish gelatin coating enriched with CUR/βCD emulsion on grass carp (*Ctenopharyngodon idellus*) fillets during storage at 4 °C. *Food Chem.* **2019**, *272*, 643–652. [CrossRef]
65. Guerra, N.P.; Macias, C.L.; Agrasar, A.T.; Castro, L.P. Development of a bioactive packaging cellophane using NisaplinR as biopreservative agent. *Lett. Appl. Microbiol.* **2005**, *40*, 106–110. [CrossRef] [PubMed]
66. Reygaert, W.C. The antimicrobial possibilities of green tea. *Front. Microbiol.* **2014**, *5*, 434. [CrossRef] [PubMed]

Article

Sensory Attributes, Microbial Activity, Fatty Acid Composition and Meat Quality Traits of Hanwoo Cattle Fed a Diet Supplemented with Stevioside and Organic Selenium

Yong Geum Shin, Dhanushka Rathnayake, Hong Seok Mun, Muhammad Ammar Dilawar, Sreynak Pov and Chul Ju Yang *

Animal Nutrition and Feed Science Laboratory, Department of Animal Science and Technology, Sunchon National University, Suncheon 57922, Korea; shin0048@naver.com (Y.G.S.); dhanus871@gmail.com (D.R.); mhs88828@nate.com (H.S.M.); ammar_dilawar@yahoo.com (M.A.D.); sreynakpov@gmail.com (S.P.)
* Correspondence: yangcj@scnu.kr; Tel.: +82-61-750-3235

Abstract: This study examined the effects of stevioside (S) and organic selenium (O-Se) supplementation on the sensory attributes, microbial activity, fatty acid composition, and meat quality traits of Hanwoo cattle (Korean native cattle). Twenty-four Hanwoo cattle (663 ± 22 kg body weight) were assigned to two dietary treatments for 8 months: control diet and 1% stevioside with 0.08% organic selenium supplemented diet. S and O-Se inclusion in the diet enhanced the final body weight, weight gain, and carcass crude protein ($p < 0.05$). Moreover, supplementation with S and O-Se had a significant effect on lowering the drip loss and shear force and enhanced the a^* (redness) of the *longissimus dorsi* muscle ($p < 0.05$). The inclusion of dietary S and O-Se improved the sum of the polyunsaturated fatty acid (ΣPUFAs) content of the meat, and the oxidative status (TBARS) values during second week of storage decreased by 42% ($p < 0.05$). On the other hand, the microbial count tended to decrease (7.62 vs. 7.41 $\log_{10}$ CFU), but it was not significant ($p > 0.05$), and all sensory attributes were enhanced in the S and O-Se supplemented diet. Overall, these results suggest that supplementation of the ruminant diet with stevioside and organic selenium improves the growth performance, carcass traits, and meat quality with enriched PUFAs profile and retards the lipid oxidation during the storage period in beef.

Keywords: stevioside; organic selenium; Hanwoo cattle; fatty acid profile; oxidative status; sensory attributes

1. Introduction

High meat quality characteristics of Hanwoo beef enhanced the consumer's preference compared to the other imported beef, such as Holstein steer and Australian Angus [1]. Moreover, lower subcutaneous fat contents and higher ossification scores, marbling scores, and loin protein content were also observed in the Hanwoo carcasses compared to Australian Angus carcasses [2]. Since dietary management has an important role in the beef industry, the implementation of promising feeding strategies and supplementation with additives is an essential factor that can determine the physiochemical qualities of the meat. Natural and healthy feed additives have gained particular interest in the global livestock sector since the ban on antibiotic growth promoters (AGP) in animal feed [3].

Many beneficial organic bioactive molecules are derived from plants because of their normal metabolic reactions [4]. The mode of action of plant-based feed additives is increasing the digestibility and absorption by modulating the beneficial intestinal microbiota [5]. The Stevia genus comprises approximately 200 species of various herbs and shrubs [6]. *Stevia rebaudiana* Bertoni is an herbaceous perennial plant belonging to the *Asteraceae* family that is native to Paraguay but has also been cultivated in China, Taiwan,

Korea, Malaysia, Canada, the United States, and some European countries. Stevioside (13-[2-O-beta-D-glucoprransyl-alpha-D-glucopy-ranosyl)oxy]kaur-16-en-18-oic acid-beta-D-glucopyranosyl) is the prominent steviol glycoside obtained from the leaves of *S. rebaudiana*; it is noncaloric and is stable at high temperatures over a wide range of pH [7]. Previous studies [8,9] reported that the stevioside exhibits numerous immunoregulation activities and has antibacterial properties, anti-inflammation properties, and maintenance of the blood lipid content. Previous studies showed that the stevia extract contains amino acids, polyphenols, and flavonoids and has antioxidant properties [10–13]. A few studies have been conducted on nonruminants, including pigs and poultry [14,15]. *S. rebaudiana* has been used as a sweetener and a feed additive because of the presence of various bioactive compounds.

A selenium deficiency has negative impacts on the health of animals and humans. Therefore, selenium is considered a vital component in diets. [16]. Selenium acts as an antioxidant against reactive oxygen species (ROS) through the glutathione peroxidase activity, a vital enzyme in the detoxification process [17,18]. Organic selenium is absorbed in the GI tract through active transmission and during the protein synthesis process and is deposited in tissues to fulfill the selenium requirements in organs and tissues [19]. In contrast, the utilization of organic selenium is more beneficial than inorganic selenium because of lower excretion via feces and urines. The antioxidant activity of selenium has been investigated in cattle [20,21], pigs [22,23], and poultry [24]. Nevertheless, no study has evaluated the combination of both stevioside (S) and organic selenium (O-Se) in the livestock sector.

This study examined the effects of S and O-Se as feed additives on the sensory attributes, fatty acid profiles, microbial activity, and meat quality traits of Hanwoo cattle.

2. Materials and Methods

2.1. Animal Ethics

The experimental protocol was approved by the Ministry for Agriculture, Forestry and Fishery in Korea, 2008 (SCNU-2017-1102).

2.2. Management of the Animals, Diets, and Experimental Design

A feeding trial was conducted for 8 months at the Animal Experimental Station, Sunchon National University, Suncheon, Korea. Briefly, 24 Korean native cattle (Hanwoo), aged 26 months and weighing approximately 663 ± 22 kg initial body weight, were enrolled in a completely randomized design. All animals were housed individually in raised cages maintained under environmentally controlled conditions, with an ambient temperature of 30 °C and average relative humidity of 70%. The cages were designed to enable the separate collection of feces and urine. At the end of the experimental period, all animals were slaughtered at the local abattoir to assess the carcass traits and meat quality parameters.

The animals were allotted randomly to two dietary treatments, with 12 animals per treatment group: control (basal diet) and treatment diet (basal diet + 1% stevioside, 0.08% organic selenium). A commercially available total mixed ration was used as the basal diet, and each Hanwoo animal was fed 12 kg per day. The treatment diet was prepared separately by incorporating 1% stevioside and 0.08% organic selenium on a weight: weight ratio basis. Table 1 lists the ingredients and chemical composition of the basal diets. The cattle were fed twice daily, divided into two feeding times (9:00 a.m. and 6:00 p.m.). The animals were given access to water ad libitum, and ventilation, lighting, and other management practices were implemented according to general practices. The CTC Bio Tech. Co. Ltd. Company, Seoul, Korea, provided the white-colored stevioside powder extracted from the *Stevia rebaudiana* leaves and organic Se. The purity of the extract was evaluated by high-performance liquid chromatography (HPLC) analysis of the stevioside sample and was determined to be 97%.

Table 1. Feed ingredients and chemical composition of experimental diet for Hanwoo cattle.

Composition	Amount
Ingredient (%, as-fed basis)	
Corn grain	43.28
Corn gluten feed	11.81
Wheat	10.47
Palm kernel expeller	9.03
Coconut meal	8.29
Lupin	5.91
Tapioca	3.41
Molasses	3.05
Rapeseed meal	2.06
Wheat flour	1.97
Soybean meal	0.50
Vitamin mineral premix [A]	0.22
Chemical composition (%DM)	
Total digestible nutrients (TDN)	74.12
Crude protein	12.50
Crude fat	3.61
Crude fiber	6.70
Crude ash	5.48
Calcium	0.84
Phosphorus	0.36
Neutral detergent fiber (NDF)	21.70
Acid detergent fiber (ADF)	10.70

[A] Premix providing following nutrients per kg of diet: vitamin (Vit.) A, 9,000,000 IU; Vit. D$_3$, 2,100,000 IU; Vit. E, 15,000 IU; Vit. K, 2000 mg; Vit. B$_1$, 1500 mg; Vit. B$_2$, 4000 mg; Vit. B$_6$, 3000 mg; Vit. B$_{12}$, 15 mg; Pan acid-Ca, 8500 mg; niacin, 20,000 mg; biotin, 110 mg; folic acid, 600 mg; Co, 300 mg; Cu, 3500 mg; Mn, 55,000 mg; Zn, 40,000 mg; I, 600 mg; Se, 130 mg.

2.3. Growth Performances and Slaughtering

The body-weight gain was measured as the difference between the initial and final body weights. The initial body weight was measured at study commencement. The subsequent average weight gain (AWG) of Hanwoo cattle during the experimental period was determined using the live weight obtained monthly. The feed intake was measured daily. At the end of the experiment (8 months), all animals were slaughtered after 24 h of starvation. The animals were stunned, and the carcasses were exsanguinated and immediately eviscerated. Approximately 2.5 cm steaks of *Longissimus dorsi* muscle were obtained from the 13th rib. For each subsequent analysis, triplicate samples were obtained from each carcass after being stored at 4 °C for 24 h in a chilling room, as reported by Bostami et al. [25].

2.4. Proximate Composition, Carcass Traits, and Cholesterol Analysis

The proximate compositions of the muscles, fat, and connective tissues were determined by removing them manually and ground. An Ultra-Turrax homogenizer (IKA Werke, GMBH & Co. KG, Staufen, Germany) was used for the homogenization process. The moisture (930.15), crude protein (990.03), crude fat (991.36), and crude ash (942.05) compositions were evaluated according to the guidelines set up of AOAC [26].

The meat quality grade was scored as 1^{++}, 1^{+}, 1, 2, and 3, according to the Korean beef quality grading system [27]. The main evaluated parameters were the marbling score, meat color, fat color, texture, and maturity. The marbling score was determined on a 7-point scale (7 = abundant and 1 = trace). A 7-point scale was used for scoring meat color (7 = dark red and 1 = bright red), and fat color (7 = yellowish and 1 = creamy white). A scale of 1 to 3 was used to score the texture (1 = firm and 3 = soft) and maturity (3 = mature and 1 = youthful).

The cholesterol content of the *Longissimus dorsi* muscle was determined using the method described by King et al. [28]. Briefly, 5 g of the meat sample was saponified using a chloroform and methanol mixture (2:1 vol:vol) [29]. The saponified samples were analyzed by gas chromatography (GC, DS 6200, Donam Co., Seongnam, Gyeonggi-do, Korea) with a flame ionization detector and a Hewlett Packard HP-5 capillary column (J and W Scientific, Folsom, CA, USA) with a 0.32 internal diameter, 30 m length, and 0.25 µm polyethylene glycol-film thickness. The initial temperature of the setup was 250 °C for 2 min and was increased gradually to 290 °C at a rate of 15 °C/min. The final temperature was increased to 310 °C at 10 °C/min, and held at that temperature for 10 min. The other chromatographic conditions were as follows: injector and detector temperatures of 280 °C, split ratio of 50:1, and injected sample volume of 2 µL.

2.5. Meat Quality Analysis (Meat Color, Drip Loss, Cooking Loss, Water Holding Capacity (WHC), and Shear Force)

The meat color of the meat samples (in triplicate) was measured using a Chroma meter (Model CR-410, Konica Minolta Sensing Inc., Osaka, Japan). According to the Commission International de l'Eclairage (CIE) system, the color was classified by the CIE L^* (lightness), CIE a^* (redness), and CIE b^* (yellowness) values. Cooking loss (in triplicate) is expressed as the percentage of weight loss and was evaluated by placing 1.5 cm-thick steaks of about 80 g in a polythene zipper bag, heating them in a water bath at 75 °C for 30 min, cooling them to room temperature, and holding them for 30 min. The cooked samples were cut (0.5 cm × 4.0 cm). The shear force was determined in each cooked meat sample using a Warner-Bratzler shear blade set (Lloyd Instruments Ltd., Hampshire, UK) by applying the following operating parameters: load cell of 50 kg, cross-head speed of 200 mm/min, and trigger force of 0.01 kgf. The drip loss (in triplicate) was determined as the weight loss during the suspension of a standardized sample (2 × 2 × 1 cm) sealed in a polythene bag at 4 °C after 7 days of storage. The water holding capacity (WHC) was determined using the method described by Grau et al. [30]. Briefly, 300 mg of the *Longissimus dorsi* muscle was placed in a filter-press device and compressed for approximately 2 min. The WHC was then calculated from triplicate meat samples, as a ratio of the meat film area to the total area using an area-line meter (Super PLANIX-a, Tamaya Technics Inc., Tokyo, Japan).

2.6. pH Value and Oxidative Stability of Meat

To determine the pH, approximately 5 g of *Longissimus dorsi* muscle (in triplicate) was cut into small pieces and homogenized with 45 mL distilled water for 60 s in an Ultra-Turrax (Janke and Kunkel, T25, Staufen Germany). Immediately after homogenization, the pH was measured using a digital pH meter (Docu-pH + meter, Sartorius, Columbus, OH, USA).

The thiobarbituric acid reactive substances (TBARS) were evaluated (in triplicate) using the procedure described by Witte et al. [31]. Briefly, 5 g of meat samples were mixed with 25 mL of 20% trichloroacetic acid (TCA) and homogenized for 30 s. Distilled water was added to prepare 50 mL of homogenate samples for centrifugation (3000× g, 4 °C, and 10 min). The supernatant was filtered through filter paper (Hyundai Co., Ltd., Seoul, Korea). Then, 5 mL of the filtrate was kept at room temperature for 15 h, and the absorbance was determined using a UV/VIS spectrophotometer (M2e, Molecular Devices, Sunnyvale, CA, USA). The TBARS value is expressed as micromoles of MDA/kg of meat.

2.7. Fatty Acid Analysis

The fatty acid composition of *longissimus dorsi* muscle samples (in triplicate) was evaluated using a direct method for the fatty acid methyl ester (FAME), as suggested by O'Fallon et al. [32] with a slight modification. Briefly, 1 g of minced meat sample was placed into a 15 mL Falcon tube and 0.7 mL of 10 N KOH in water and 6.3 mL of methanol were added. The tube was kept in a water bath (55 °C; 1.5 h), allowing proper permeation, hydrolyzation, and dissolution by vigorous shaking for 10 s every 30 min. After placing in a cold tap water bath, 0.58 mL of 24 N H_2SO_4 was added to precipitate K_2SO_4. The precipitated sample was placed again in a water bath (55 °C; 1.5 h) with strong shaking for 10 s every 30 min. After FAME synthesis, 3 mL of hexane was added and subjected to centrifugation for 5 min at 3000 rpm (Hanil, Combi-514R, Gimpo, Korea). The top hexane layer, which contained FAME, was dehydrated by passing through the anhydrous Na_2SO_4. The extracted and dehydrated hexane was placed into a GC vial and concentrated to 1.5 mL for analysis.

The fatty acid composition of the FAME was determined using an Agilent gas chromatography system (6890 N, Agilent Technologies, Santa Clara, CA, USA). Briefly, fat was extracted from minced meat using a chloroform-methanol (2:1 v/v) solution [28]. According to the AOAC [26] procedure, the prepared fatty acid methyl esters were dissolved in hexane before injection; 1 µL of the prepared sample was injected into the GC; the set-up injector temperature was maintained at 250 °C with a 100:1 split ratio, using a WCOT-fused silica capillary column (100 m × 0.25 m i.d., 0.20 µm film thickness; Varian Inc., Palo Alto, CA, USA) with helium flow. The oven conditions were 150 °C/1 min, 150–200 °C at 7 °C/min, 200 °C/5 min, 200–250 °C at 5 °C/min, and 250 °C/10 min. The setup detector temperature was 275 °C. Fatty acid peaks were evaluated using the retention time of fatty acid standards (47015-U, Sigma-Aldrich Corp., LLC., St. Lois, MO, USA). The proportion (%) against the total peak area was calculated from the peak area of each fatty acid identified.

2.8. Meat Microbial Analysis and Sensory Evaluation

Triplicates of *longissimus dorsi* muscle from each group were taken for the meat microbial analysis. A 25 g sample of meat was homogenized using 225 mL of a NaCl solution (0.85% W/V). Subsequently, 20 µL was obtained from 10-fold diluted solution and transferred into tryptic soy agar plates (Becton, Dickinson, and Company, Sparks, MD 21152, USA) using a sterilized triangle spreader for microbial enumeration. The colonies were counted immediately after incubation. The microbial number was determined as follows: number of colonies × 10 dilution value × (100/20) = multiplied value = log (multiplied value). The ultimate count was expressed as log_{10} CFU/g.

Ten samples from both treatment groups were examined by 10 members, all well-trained experts of the sensory panel at the Department of Animal Science and Technology, Sunchon National University, Suncheon, Korea. Moreover, individual testing booths and a controlled lighting facility were provided during the sensory evaluation process [33]. Each steak was cooked at approximately 150 °C and until the internal temperature reached 70 °C, which was determined by inserting a digital thermometer in the steak. Before evaluation by the panelists, the steaks were wrapped in an aluminum foil and kept under 65 °C in an oven. The flavor, tenderness, and juiciness were then evaluated by the panelists. A 7-point hedonic scale was used to express the value of each characteristic: 7 indicated desirable flavor, extremely tender, reddish color, high palatability, and good juiciness, while 1 was indicative of undesirable favor, extremely tough, pale color, low palatability, and extremely dry. The average value of the 10 panelists determined the color, flavor, tenderness, juiciness, and palatability of the cut.

2.9. Statistical Analysis

The data were analyzed using the General Linear Model (GLM) method of the Statistics Package Program (SAS, 2003, Version 9.1, SAS Institute, Cary, NC, USA). For the growth performance parameters, a group of two cattle served as the experimental unit. Individual animals served as the experimental unit for the carcass traits, the meat quality parameters, cholesterol content, fatty acid profile, pH, TBARS, microbial analysis, and sensory evaluation.

$$Y_{ij} = \mu + \alpha_i + e_{ij}. \tag{1}$$

where Y_{ij} is the response variable, μ is the general mean value, α_i is the effect of dietary supplementation, and e_{ij} is the random error. The mean values were compared using a student's *t*-test. The level of significance considered for the tests was $p < 0.05$.

3. Results

3.1. Growth Performance, Proximate Analysis, and Cholesterol Content

Over the entire experimental period, the final body weight and body weight gain were significantly higher ($p < 0.05$) in the S and O-Se supplemented diet. On the other hand, the feed intake did not have a significant effect but was higher in the S- and O-Se-supplemented diet (Table 2).

Table 2. Effect of stevioside and organic selenium on growth performances of Hanwoo cattle (experimental period: 243 days).

Item	CON [1]	TRT [2]	SEM	p-Value
Initial BW	647.00	652.00	22.01	0.32
Final BW	717.79 [b]	751.93 [a]	31.47	0.04
Weight gain (kg)	0.30 [b]	0.42 [a]	0.14	0.03
Feed intake (kg/Day)	11.32	11.43	0.22	0.51

[a,b] Means in the same row with different superscripts are significantly different; SEM = standard error of mean. [1] CON: control diet with no added stevioside and organic selenium; [2] TRT: treatment (basal diet + 1% stevioside + 0.08% organic selenium).

Proximate analysis revealed a significantly higher crude protein content, lower crude fat ($p < 0.05$), and a numerically lower meat cholesterol amount due to S and O-Se supplementation even if it is not significantly different (Table 3).

Table 3. Effect of stevioside and organic selenium on proximate analysis and cholesterol contents in the meats of Hanwoo cattle.

Item	CON [1]	TRT [2]	SEM	p-Value
Moisture (%)	56.83 [b]	63.89 [a]	1.23	0.01
Crude protein (%)	22.48 [b]	25.26 [a]	0.80	0.03
Crude fat (%)	16.62 [a]	12.45 [b]	0.95	0.01
Crude ash (%)	1.18	1.29	0.05	0.14
Cholesterol (mg/100 g)	50.27	49.74	0.04	0.92

[a,b] Means in the same row with different superscripts are significantly different; SEM = standard error of mean; [1] CON: control diet with no added stevioside and organic selenium; [2] TRT: treatment (basal diet + 1% stevioside + 0.08% organic selenium).

3.2. Carcass Traits

Although the S- and O-Se-supplemented diet increases the carcass yield, back-fat thickness, and loin area, the changes were not significant ($p > 0.05$). Moreover, the changes in the meat quality were also not significant, including the marbling score, meat color, fat color, texture, and maturity (Table 4).

3.3. Meat Quality Analysis (Meat Color, Drip Loss, Cooking Loss, WHC, and Shear Force)

Considering the changes in the meat color, the redness (a^*) was enhanced significantly in the S- and O-Se-supplemented diet ($p < 0.05$). On the other hand, no significant difference in surface lightness (L^*) and yellowness (y^*) was observed between the two treatments. Although S and O-Se addition did not affect the cooking loss and water holding capacity (WHC) ($p > 0.05$), the drip loss and shear force decreased significantly ($p < 0.05$) (Table 5).

3.4. pH Value and Oxidative Stability of Meat

The pH in both treatments decreased gradually up to the second week of storage, with a subsequent increase at the end of the third week. On the other hand, no significant difference was observed between the control and S- and O-Se-added diet (Figure 1). The addition of S and O-Se resulted in significantly lower TBARS values during second week of storage ($p < 0.05$) (Figure 2).

Table 4. Effect of stevioside and organic selenium supplementation on carcass traits of Hanwoo cattle.

Items	CON [1]	TRT [2]	SEM	p-Value
Yield traits				
Carcass weight (kg)	446.68	449.57	20.46	0.87
Loin area (cm^2)	100.50	103.33	4.42	0.67
Back fat thickness (mm)	16.67	18.33	2.81	0.68
Quality traits				
Marbling score [3]	4.83	6.33	0.78	0.20
Meat color [4]	4.83	4.83	0.17	1.00
Fat color [5]	3.17	2.83	0.19	0.18
Texture [6]	1.17	1.00	0.08	0.34
Maturity [7]	2.00	1.89	0.16	0.21
Quality grade	0:3:1:2:0	2:1:3:0:0	–	–
Meat point [8]	3.17	3.83	0.40	0.26

SEM = standard error of mean; [1] CON: control diet with no added stevioside and organic selenium; [2] TRT: treatment (basal diet + 1% stevioside + 0.08% organic selenium); [3] Marbling score: 7 = abundant, 1 = trace; [4] meat color: 1 = bright red and 7 = dark red; [5] fat color: 1 = creamy white and 7 = yellowish.; [6] texture: 1 = firm and 3 = soft.; [7] maturity: 1 = young and 3 = youthful; [8] meat point: 1++: 5 point, 1+: 4 Point, 1: 3 Point, 2: 2 Point, and 3: 1 Point.

Table 5. Effect of stevioside and organic selenium on meat quality of *Longissimus dorsi* muscle from Hanwoo cattle.

Item	CON [1]	TRT [2]	SEM	p-Value
Meat color				
CIE L^*	29.04	29.21	0.73	0.87
CIE a^*	14.94 [b]	18.45 [a]	0.81	0.03
CIE b^*	3.58	5.87	0.70	0.07
Drip loss (%)	21.14 [a]	16.48 [b]	0.75	0.00
Cooking loss (%)	15.81	16.30	0.71	0.70
WHC (%)	12.81	13.30	1.26	0.78
Shear force (kg)	5.38 [a]	3.93 [b]	0.35	0.01

[a,b] Means in the same row with different superscripts are significantly different; SEM = standard error of mean; [1] CON: control diet with no added stevioside and organic selenium; [2] TRT: treatment (basal diet + 1% stevioside + 0.08% organic selenium).

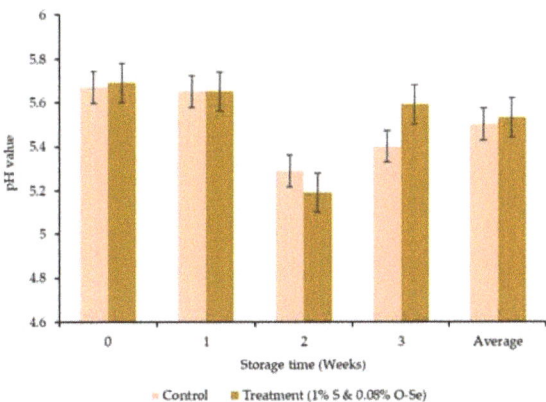

Figure 1. Effect of stevioside and organic selenium on the pH values in meats of Hanwoo cattle.

3.5. Fatty Acid Analysis

Monounsaturated fatty acids (MUFA) had the highest proportion. They constituted more than 50% of the total fatty acid composition, followed by saturated fatty acids (SFA) and polyunsaturated fatty acids (PUFA), accounting for 37–39% and 0.6–0.9%, respectively. Oleic acid (C8:1, 47%) was the predominant fatty acid among MUFA, followed by palmitoleic acid (C16:1, 5%), while palmitic acid (C16:0, 24–26%), stearic acid (C18:0, 8%) and myristic acid (C14:0, 3%) were predominant among SFA. On the other hand, the decrease in SFAs content was not significant in the S- and O-Se-supplemented diet. Moreover, the nervonic monounsaturated fatty acid content and linoleic, di homo-gamma linolenic, and arachidonic PUFAs contents were also higher in the animals fed the S- and O-Se-supplemented diet ($p < 0.05$) (Table 6).

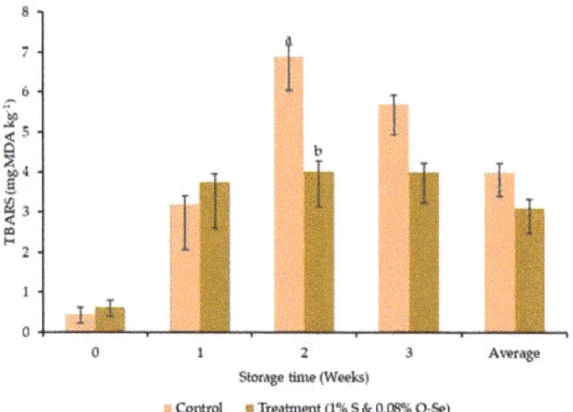

Figure 2. Effect of stevioside and organic selenium on the TBARS values in meats of Hanwoo cattle. Data presented as the mean ± s.e. bars at a specific time point with different letters show a significant difference ($p < 0.05$).

Table 6. Fatty acid profile of *Longissimus dorsi* muscle from Hanwoo cattle fed diets containing stevioside and organic selenium.

Items	CON [1]	TRT [2]	SEM	p-Value
C10:0	0.03	0.04	0.00	0.08
C12:0	0.08	0.08	0.00	0.69
C14:0	3.35	3.39	0.11	0.80
C15:0	0.20 [b]	0.25 [a]	0.01	0.008
C16:0	26.65 [a]	24.05 [b]	0.60	0.04
C17:0	0.92 [a]	0.15 [b]	0.03	<0.0001
C18:0	8.33	8.67	0.15	0.13
C20:0	0.08 [a]	0.07 [b]	0.00	0.004
C14:1	0.30 [a]	0.25 [b]	0.02	0.04
C16:1	5.92 [a]	5.62 [b]	0.06	0.007
C18:1	47.19	46.09	0.58	0.24
C24:1	0.06 [b]	0.08 [a]	0.00	0.01
C18:2	0.06 [b]	0.10 [a]	0.01	0.004
C18:3	0.22 [b]	0.36 [a]	0.04	0.008
C20:3	0.13 [b]	0.16 [a]	0.01	0.04
C20:4	0.20 [b]	0.26 [a]	0.01	0.03
ΣSFA [3]	39.69	36.75	0.72	0.16
ΣMUFA [4]	53.46	52.04	0.58	0.13
ΣPUFA [5]	0.63 [b]	0.88 [a]	0.05	0.04
ΣPUFA/SFA	0.015	0.024	0.04	0.14

[a,b] Means in the same row with different superscripts are significantly different; SEM = standard error of mean; [1] CON: control diet with no added stevioside and organic selenium; [2] TRT: treatment (basal diet + 1% stevioside supplementation + 0.08% organic selenium); [3] SFA: saturated fatty acid; [4] MUFA: monounsaturated fatty acid; [5] PUFA: polyunsaturated fatty acid.

3.6. Microbial Analysis and Sensory Evaluation

The average meat microbial content was numerically lower, but not significant (7.62 vs. 7.41 $\log_{10}$ CFU) in the S- and O-Se-supplemented diet compared to the control treatment after the 3-week refrigerated storage period (Figure 3).

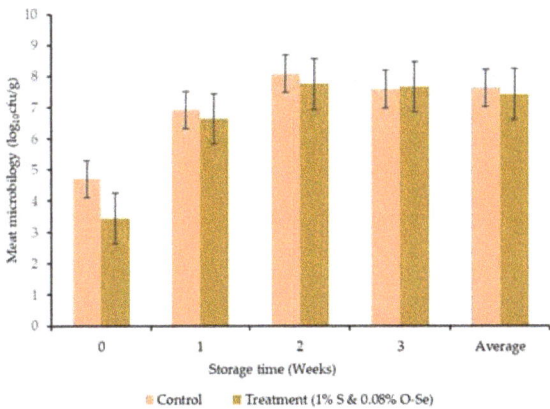

Figure 3. Effect of stevioside and organic selenium on the microbial content in meats of Hanwoo cattle.

Although no significant impact ($p > 0.05$) was observed for color, flavor, tenderness, juiciness, and palatability, the S- and O-Se-supplemented diet had a positive tendency on the sensory parameters (Figure 4).

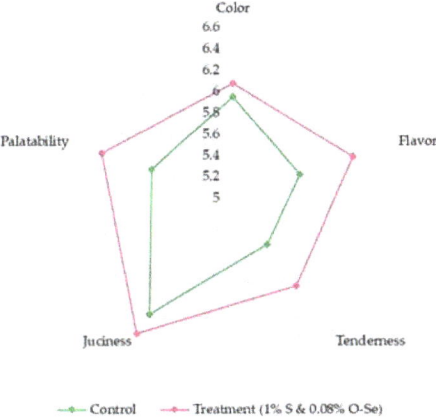

Figure 4. Effect of stevioside and organic selenium on meat sensory attributes of Hanwoo cattle.

4. Discussion

The incorporation of feed additives has gained popularity because of their benefits to human and animal health [34,35]. In particular, phytochemicals have biochemical properties, including antioxidant, antimicrobial, antistress, and nutrigenomic influences on the development of immunity and improved productivity [36,37]. Moreover, the incorporation of organic selenium can improve nutrient utilization by preventing the pro-oxidant effects on the gut system and has an antioxidant defense role against oxidative stress [38].

In the present study, 1% stevioside and 0.08% organic selenium supplementation significantly increased the weight gain and final body weight and enhanced the feed intake (FI). The palatability of the feed was enhanced because of the sweetness of the stevioside, and the animals increased their voluntary feed intake, resulting in increased body weight. In contrast, stevioside derivatives can stimulate the taste receptors (TASIR2/TASIR3) [39,40] and K^+ channels in pancreatic β-cells [41], which can increase the stimulation of appetite and voluntary FI. Similar to the present findings, Han et al. 2019 [42] reported a linear increase in FI in goats given diets supplemented with 270 and 541 mg/kg stevioside. In contrast, Cho et al. [43] reported that the dietary intake of Hanwoo steers is not influenced by the inclusion of 65 mg/kg stevioside as an essential oil to the feed with 0.1% of diet supplemented rate. Furthermore, previous studies reported improved production parameters in pigs [44], layers [45], broilers [46], and cattle [47] owing to O-Se supplementation. The enhanced production performance may be associated with the developed antioxidant capacity and decrease in microbial pathogen colonization in the gut system because of the synergistic action of S and O-Se supplementation.

No studies have evaluated the effects of the dietary inclusion of S and O-Se on the carcass traits, meat quality, and sensory evaluation of cattle. A Se deficiency increases the susceptibility to various degenerative diseases in humans [48]. The incorporation of Se into animal feed has great potential to obtain Se-enriched meat that can alleviate Se deficiencies [49]. Dietary S and O-Se supplementation increased the moisture and crude protein content with a concomitant reduction in the crude fat content in meat taken from Hanwoo cattle. The lower crude fat content in meat might be associated with the lipolytic mechanism owing to the presence of polyphenols and flavonoids compounds in *S. rebaudiana* [50]. The increase in moisture content is caused by the inverse relationship between the meat fat and moisture content, which affects the meat tenderness and juiciness [51]. The cholesterol content was lower in the S and O-Se-included treatment, probably due to the cooperative mechanism of flavonoids in stevioside and glutathione peroxidase activity in organic Se [52], which can convert cholesterol to bile acids through the induced stimulation of enzymes activity. Consequently, it is catabolized and eliminated from the body.

Genetics and environmental factors, including the feeding strategies, affect the carcass quality traits directly [53]. Although no significant difference was observed in back-fat thickness of both treatments, Zhang et al. [54] stated that O-Se can provide some vitamins, amino acids, protein, and other nutrients for the body and can cause subsequent increase in fat accumulation. Similar to that, Choi et al. [55] also reported significant differences in the carcass length and back-fat thickness of pork meat obtained from animals fed stevia and charcoal-supplemented diets. Nevertheless, 0.3% stevia and charcoal inclusion did not significantly enhance the marbling, firmness, and color scores in pigs compared to the control treatment. Therefore, further investigations will be needed to determine the impact on the carcass traits based on different concentrations of stevioside and organic Se application.

The meat color and brightness are considered important visual factors that affect the consumer's preference and purchasing decisions [56]. In the current study, the redness (a^*) value was enhanced by the inclusion of S and O-Se in the cattle diet. The increase in meat redness may be associated with the reduced metmyoglobin (MMG) formation by antioxidants in both S and O-Se by delaying meat oxymyoglobin (OMG) oxidation. Choi et al. [55] also reported higher L^*, a^*, and b^* values in pork meat supplemented with 0.3% stevioside and charcoal. The WHC was not significant in the two treatments, but the drip loss and shear force values were significantly lower in the S- and O-Se-supplemented group. The meat pH and lipid peroxidation affected meat drip loss [57]. Therefore, in the current study, the reduced lipid oxidation caused by S and O-Se supplementation may affect the lower drip loss value. Moreover, the S and O-Se supplemented diet had no adverse effect on the meat quality parameters.

High accumulated lactic acid content due to the metabolism of the glycogen reserves and releasing of H^+ from ATP hydrolysis will lead lower muscle pH. Therefore, the carcass pH has a significant influence on the meat quality traits. During the 3 weeks in the S- and O-Se-included diet, the high pH of the meat probably affected the glutathione peroxidase enzyme activity, which can break down H_2O_2 to H_2O and O_2 [58] and facilitate an increase in meat pH. Moreover, in the current study, the TBARS value during the second week of the storage periods was 42% lower in the S- and O-Se-supplemented diet group than the control group. In addition to acting as a sweetener, previous studies reported the antioxidant properties of stevioside [59–61]. Moreover, plant polyphenolic compounds can prevent the oxidation activities in unsaturated fatty acids by scavenging free radicals [62] or singlet-oxygen quenching ability [63]. The glutathione peroxidase family (GPx) in organic Se can also delay the lipid oxidation reactions [64]. Overall, the synergistic effects of both S and O-Se reduced the average TBARS value owing to the presence of antioxidants.

The meat fatty acids composition depends on the lipogenesis activities in adipose tissues and the ruminal biohydrogenation process [65]. The dietary SFA has a direct correlation with cholesterol level and subsequent cardiovascular diseases. Nevertheless, PUFA positively exerts a lower cholesterol content and a significant reduction in human health risks [66]. In the present study, the PUFA content was significantly higher in the S- and O-Se-supplemented diet. This contrasts with a previous study [67], which noted a higher linoleic content in fattening steers fed a diet containing O-Se, possibly because of a reduction in biohydrogenation process and subsequent increase in the intestinal absorption of PUFA. A previous study [24] reported a lower MUFA and a higher PUFA proportion in the O-Se-supplemented group owing to the presence of antioxidants. Moreover, phytochemicals in plants indirectly performed their antioxidants activities against the depletion of PUFA from microbial biohydrogenation and enhanced the UFA concentration in the muscles [68]. Therefore, the combination of S and O-Se has a positive impact on the PUFA content in the meat.

A combination of microbes and endogenous enzymes leads to the deterioration of meat protein and consequently accelerates meat spoilage [69]. Synthetic preservatives are the main approach to inhibit microbial growth. On the other hand, preservatives need to be replaced by harmless natural compounds to avoid harmful effects on human health [70]. In the present study, the average microbial count tended to decrease in S and O-Se treatment during the refrigerated storage period, possibly because the GSH-Px enzymatic activity can eradicate harmful lipid peroxide and H_2O_2 from organisms, resulting in less favorable reproductive conditions for microbes. Hence, the microbial population tends to decrease gradually [54]. Moreover, antioxidant molecules in stevioside can scavenge free radicals that influence the retarded muscle oxidation process and reduce the microbial flora content in meat. In our study, the supplementation of S and O-Se did not modify the meat microbial count significantly. Hence, the incorporation of different concentrations of S and O-Se in cattle diets requires further study.

The sensory evaluation helps determine the consumer preference or acceptability of meat through an evaluation via sight, aroma, taste, and touch. In the present study, the supplementation of S and O-Se tend to positively influence color, flavor, tenderness, juiciness, and palatability of meat. Lipid peroxidation influences the deterioration of the above sensory properties of meat and the optimal quality of meat products [54,71]. Therefore, the antioxidant properties of S [72] and O-Se probably helps reduce the lipid oxidation process of meat and helps enhance the sensory properties. Interestingly, Zou et al. [70] reported that O-Se could retard the oxidation of OMG and lipids, resulting in a higher meat color score and meat stability. Furthermore, previous studies confirmed that the O-Se-incorporated diet in the animal feed improves the meat quality by preventing excessive dehydration and protects the meat sensory attributes [73]. Consistently, this study indicated that the synergism of S and O-Se in the cattle diet led to an improvement in the meat sensory attributes.

5. Conclusions

The addition of S and O-Se in the cattle diet enhanced the final body weight (FBW) and weight gain (WG). In addition, the carcass protein and moisture contents were improved, and the cholesterol content was reduced. Both S and O-Se inclusion improved the meat redness ($a*$) and reduced the meat drip loss and shear force value. The average TBARS value in S and O-Se supplemented diet during the second week was decreased by 42%

and showed a higher meat ΣPUFA content than the control group. In this study, all sensory attributes tended to increase numerically and microbial flora contents also tended to decrease due to the S- and O-Se-incorporated diet. These results indicated that the supplementation of natural stevioside and organic selenium could be carried out to enhance the productive performance and the quality of meat in the livestock sector. Nevertheless, further studies will be needed to determine if different concentrations could provide more benefit in ruminant diets.

Author Contributions: Conceptualization, C.J.Y. and H.S.M.; methodology, Y.G.S. and D.R.; software, M.A.D. and S.P.; validation, C.J.Y. and H.S.M.; formal analysis, C.J.Y.; investigation, H.S.M.; resources, Y.G.S.; data curation, D.R. and S.P.; writing—original draft preparation, Y.G.S., D.R., and M.A.D.; writing—review and editing, D.R. and Y.G.S.; supervision, C.J.Y.; project administration, H.S.M.; funding acquisition, H.S.M. All authors have read and agreed to the published version of the manuscript.

Funding: This study was supported by the CTC Bio Tech. Co. Ltd. Company, Seoul, Korea (Grant No. 2016-0301).

Institutional Review Board Statement: The experimental protocol was approved by the Ministry for Agriculture, Forestry and Fishery in Korea, 2008 (SCNU-2017-1102).

Informed Consent Statement: Not applicable.

Data Availability Statement: The data presented in this study are available in the article.

Conflicts of Interest: The authors declare no conflict of interest.

References

1. Hwang, Y.H.; Kim, G.-D.; Jeong, J.Y.; Hur, S.J.; Joo, S.T. The relationship between muscle fiber characteristics and meat quality traits of highly marbled Hanwoo (Korean native cattle) steers. *Meat Sci.* **2010**, *86*, 456–461. [CrossRef] [PubMed]
2. Jo, C.; Cho, S.H.; Chang, J.; Nam, K.C. Keys to production and processing of Hanwoo beef: A perspective of tradition and science. *Anim. Front.* **2012**, *2*, 32–38. [CrossRef]
3. Kamel, C. Natural Plant Extracts: Classical Medies Bring Modern Animal Production Solutions. Available online: https://agris.fao.org/agris-search/search.do?recordID=QC2001600008 (accessed on 28 November 2008).
4. Suresh, G.; Das, R.K.; Brar, S.K.; Rouissi, T.; Ramirez, A.; Chorfi, Y.; Godbout, S. Alternatives to antibiotics in poultry feed: Molecular perspectives. *Crit. Rev. Microbiol.* **2017**, *44*, 318–355. [CrossRef]
5. Ahmed, S.T.; Hossain, M.E.; Kim, G.M.; Hwang, J.A.; Ji, H.; Yang, C.J. Effects of Resveratrol and Essential Oils on Growth Performance, Immunity, Digestibility and Fecal Microbial Shedding in Challenged Piglets. *Asian Australas. J. Anim. Sci.* **2013**, *26*, 683–690. [CrossRef] [PubMed]
6. Yadav, A.K.; Singh, S.; Dhyani, D.; Ahuja, P.S. A review on the improvement of stevia [*Stevia rebaudiana* (Bertoni)]. *Can. J. Plant Sci.* **2011**, *91*, 1–27. [CrossRef]
7. Puri, M.; Sharma, D.; Tiwari, A.K. Downstream processing of stevioside and its potential applications. *Biotechnol. Adv.* **2011**, *29*, 781–791. [CrossRef]
8. Takasaki, M.; Konoshima, T.; Kozuka, M.; Tokuda, H.; Takayasu, J.; Nishino, H.; Miyakoshi, M.; Mizutani, K.; Lee, K.-H. Cancer preventive agents. Part 8: Chemopreventive effects of stevioside and related compounds. *Bioorganic Med. Chem.* **2009**, *17*, 600–605. [CrossRef]
9. Pariwat, P.; Homvisasevongsa, S.; Muanprasat, C.; Chatsudthipong, V. A Natural Plant-Derived Dihydroisosteviol Prevents Cholera Toxin-Induced Intestinal Fluid Secretion. *J. Pharmacol. Exp. Ther.* **2008**, *324*, 798–805. [CrossRef]
10. Abou-Arab, A.E.; Abou-Arab, A.A.; Abu-Salem, M.F. Physico-chemical assessment of natural sweeteners steviosides produced from *Stevia rebaudiana* Bertoni plant. *Afr. J. Food Sci.* **2010**, *4*, 269–281. [CrossRef]
11. Muanda, F.N.; Soulimani, R.; Diop, B.; Dicko, A. Study on chemical composition and biological activities of essential oil and extracts from *Stevia rebaudiana* Bertoni leaves. *LWT Food Sci. Technol.* **2011**, *44*, 1865–1872. [CrossRef]
12. Shukla, S.; Mehta, A.; Bajpai, V.K.; Shukla, S. In vitro antioxidant activity and total phenolic content of ethanolic leaf extract of *Stevia rebaudiana* Bert. *Food Chem. Toxicol.* **2009**, *47*, 2338–2343. [CrossRef] [PubMed]
13. Tadhani, M.B.; Patel, V.H.; Subhash, R. In vitro antioxidant activities of *Stevia rebaudiana* leaves and callus. *J. Food Compos. Anal.* **2007**, *20*, 323–329. [CrossRef]
14. Munro, P.J.; Lirette, A.; Anderson, D.M.; Ju, H.Y. Effects of a new sweetener, Stevia, on performance of newly weaned pigs. *Can. J. Anim. Sci.* **2000**, *80*, 529–531. [CrossRef]
15. Paredes-López, D.; Robles-Huaynate, R.; Carrión-Molina, M. Effect of *Stevia rebaudiana* Bertonni Leaves powder on Lipid Profiles and Productive Parameters of Laying Hens. *Sci. Agropecu* **2019**, *10*, 275–282. [CrossRef]

16. Alfthan, G.; Eurola, M.; Ekholm, P.; Venäläinen, E.-R.; Root, T.; Korkalainen, K.; Hartikainen, H.; Salminen, P.; Hietaniemi, V.; Aspila, P.; et al. Effects of nationwide addition of selenium to fertilizers on foods, and animal and human health in Finland: From deficiency to optimal selenium status of the population. *J. Trace Elem. Med. Biol.* **2015**, *31*, 142–147. [CrossRef]
17. Albanes, D.; Till, C.; Klein, E.A.; Goodman, P.J.; Mondul, A.M.; Weinstein, S.J.; Taylor, P.R.; Parnes, H.L.; Gaziano, J.M.; Song, X.; et al. Plasma Tocopherols and Risk of Prostate Cancer in the Selenium and Vitamin E Cancer Prevention Trial (SELECT). *Cancer Prev. Res.* **2014**, *7*, 886–895. [CrossRef]
18. Speckmann, B.; Grune, T. Epigenetic effects of selenium and their implications for health. *Epigenetics* **2015**, *10*, 179–190. [CrossRef]
19. Schrauzer, G.N. The nutritional significance, metabolism and toxicology of selenomethionine. *Adv. Food Nutr. Res.* **2003**, *47*, 73–112. [CrossRef]
20. Taylor, J.B.; Marchello, M.J.; Finley, J.W.; Neville, T.L.; Combs, G.F.; Caton, J.S. Nutritive value and display-life attributes of selenium-enriched beef-muscle foods. *J. Food Compos. Anal.* **2008**, *21*, 183–186. [CrossRef]
21. Cozzi, G.; Prevedello, P.; Stefani, A.L.; Piron, A.; Contiero, B.; Lante, A.; Gottardo, F.; Chevaux, E. Effect of dietary supplementation with different sources of selenium on growth response, selenium blood levels and meat quality of intensively finished Charolais young bulls. *Animal* **2011**, *5*, 1531–1538. [CrossRef] [PubMed]
22. Svoboda, M.; Saláková, A.; Fajt, Z.; Ficek, R.; Buchtová, H.; Drábek, J. Selenium from Se-enriched lactic acid bacteria as a new Se source for growing-finishing pigs. *Pol. J. Vet. Sci.* **2009**, *12*, 355–361. [PubMed]
23. Zhan, X.; Wang, M.; Zhao, R.; Li, W.; Xu, Z. Effects of different selenium source on selenium distribution, loin quality and antioxidant status in finishing pigs. *Anim. Feed Sci. Technol.* **2007**, *132*, 202–211. [CrossRef]
24. Biazik, E.; Straková, E.; Suchý, P. Effects of Organic Selenium in Broiler Feed on the Content of Selenium and Fatty Acid Profile in Lipids of Thigh Muscle Tissue. *Acta Vet. Brno* **2013**, *82*, 277–282. [CrossRef]
25. Rubayet Bostami, A.B.M.; Mun, H.S.; Yang, C.J. Loin eye muscle physico-chemical attributes, sensory evaluation and proximate composition in Korean Hanwoo cattle subjected to slaughtering along with stunning with or without pithing. *Meat Sci.* **2018**, *145*, 220–229. [CrossRef] [PubMed]
26. Association of Official Analytical Chemists (AOAC). *Official Methods of Analysis of AOAC International*, 17th ed.; AOAC International: Gaithersburg, MD, USA, 2000.
27. KAPE 2012. 'The beef carcass grading.' (Korea Institute for Animal Products Quality Evaluation, Gunposi, Gyungki-do, South Korea). Available online: http://www.ekape.or.kr/view/eng/system/beef.asp (accessed on 15 May 2012).
28. King, A.J.; Paniangvait, P.; Jones, A.D.; German, J.B. Rapid Method for Quantification of Cholesterol in Turkey Meat and Products. *J. Food Sci.* **1998**, *63*, 382–385. [CrossRef]
29. Folch, J.; Lees, M.; Stanley, G.S. A simple method for the isolation and purification of total lipids from animal tissues. *J. Biol. Chem.* **1957**, *226*, 497–509. [CrossRef]
30. Grau, R.; Hamm, G. A simple method for determining water binding in muscles. *Diet Nat.* **1953**, *40*, 29–30.
31. Witte, V.C.; Krause, G.F.; Bailey, M.E. A new extraction method for determining 2-thiobarbituric acid values of pork and beef during storage. *J. Food Sci.* **1970**, *35*, 582–585. [CrossRef]
32. O'Fallon, J.V.; Busboom, J.R.; Nelson, M.L.; Gaskins, C.T. A direct method for fatty acid methyl ester synthesis: Application to wet tissues, oils, and feedstuffs. *J. Anim. Sci.* **2007**, *85*, 1511–1521. [CrossRef]
33. Meilgaard, M.C.; Carr, B.T.; Civille, G.V. *Sensory Evaluation Techniques*; CRC Press: Boca Raton, FL, USA, 2006; pp. 25–28. ISBN 978-1-00-304072-9.
34. Geetha, V.; Chakravarthula, S.N. Chemical composition and anti-inflammatory activity of *Boswellia ovalifoliolata* essential oils from leaf and bark. *J. For. Res.* **2018**, *29*, 373–381. [CrossRef]
35. Dhama, K.; Karthik, K.; Khandia, R.; Munjal, A.; Tiwari, R.; Rana, R.; Khurana, S.K.; Sana Ullah, K.R.; Alagawany, M.; Farag, M.R.; et al. Medicinal and therapeutic potential of herbs and plant metabolites/extracts countering viral pathogens-Current knowledge and future prospects. *Curr. Drug Metab.* **2018**, *19*, 236–263. [CrossRef] [PubMed]
36. Golestan, I. Phytogenics as new class of feed additive in poultry industry. *J. Anim. Vet. Adv.* **2010**, *9*, 2295–2304.
37. Krause, D.O.; House, J.D.; Nyachoti, C.M. Alternatives to antibiotics in swine diets: A molecular approach. *Proc. Manit. Swine Sem.* **2005**, *19*, 57–66.
38. Surai, P.F.; Kochish, I.I.; Fisinin, V.I.; Velichko, O.A. Selenium in Poultry Nutrition: From Sodium Selenite to Organic Selenium Sources. *Jpn. Poult. Sci.* **2018**, *55*, 79–93. [CrossRef]
39. Hellfritsch, C.; Brockhoff, A.; Stähler, F.; Meyerhof, W.; Hofmann, T. Human Psychometric and Taste Receptor Responses to Steviol Glycosides. *J. Agric. Food Chem.* **2012**, *60*, 6782–6793. [CrossRef]
40. Ahmed, J.; Preissner, S.; Dunkel, M.; Worth, C.L.; Eckert, A.; Preissner, R. Super Sweet—A resource on natural and artificial sweetening agents. *Nucleic Acids Res.* **2010**, *14*, 39.
41. Abudula, R.; Matchkov, V.V.; Jeppesen, P.B.; Nilsson, H.; Aalkjaer, C.; Hermansen, K. Rebaudioside A directly stimulates insulin secretion from pancreatic beta cells: A glucose-dependent action via inhibition of ATP-sensitive K^+-channels*. *Diabetes Obes. Metab.* **2008**, *10*, 1074–1085. [CrossRef]
42. Han, X.; Chen, C.; Zhang, X.; Wei, Y.; Tang, S.; Wang, J.; Tan, Z.; Xu, L. Effects of Dietary Stevioside Supplementation on Feed Intake, Digestion, Ruminal Fermentation, and Blood Metabolites of Goats. *Animals* **2019**, *9*, 32. [CrossRef]

43. Cho, S.; Mbiriri, D.T.; Shim, K.; Lee, A.L.; Oh, S.J.; Yang, J.; Ryu, C.; Kim, Y.H.; Seo, K.S.; Chae, J.I.; et al. The influence of feed energy density and a formulated additive on rumen and rectal temperature in hanwoo steers. *Asian Australas. J. Anim. Sci.* **2014**, *27*, 1652–1662.
44. Quesnel, H.; Renaudin, A.; Le Floc'h, N.; Jondreville, C.; Père, M.C.; Taylor-Pickard, J.A.; Le Dividich, J. Effect of organic and inorganic selenium sources in sow diets on colostrum production and piglet response to a poor sanitary environment after weaning. *ANM* **2008**, *2*. [CrossRef] [PubMed]
45. Tufarelli, V.; Ceci, E.; Laudadio, V. 2-Hydroxy-4-Methylselenobutanoic Acid as New Organic Selenium Dietary Supplement to Produce Selenium-Enriched Eggs. *Biol. Trace Elem. Res.* **2016**, *171*, 453–458. [CrossRef] [PubMed]
46. Surai, P.F.; Fisinin, V.I. Selenium in poultry breeder nutrition: An update. *Anim. Feed Sci. Technol.* **2014**, *191*, 1–15. [CrossRef]
47. Khalili, M.; Chamani, M.; Amanlou, H.; Nikkhah, A.; Sadeghi, A. The effect of feeding inorganic and organic selenium sources on the performance and content of selenium in milk of transition dairy cows. *Acta Sci. Anim. Sci.* **2019**, *41*, 44691. [CrossRef]
48. Gromadzińska, J.; Reszka, E.; Bruzelius, K.; Wąsowicz, W.; Åkesson, B. Selenium and cancer: Biomarkers of selenium status and molecular action of selenium supplements. *Eur. J. Nutr.* **2008**, *47*, 29–50. [CrossRef]
49. Zhang, W.; Xiao, S.; Samaraweera, H.; Lee, E.J.; Ahn, D.U. Improving functional value of meat products. *Meat Sci.* **2010**, *86*, 15–31. [CrossRef]
50. Gaweł-Bęben, K.; Bujak, T.; Nizioł-Łukaszewska, Z.; Antosiewicz, B.; Jakubczyk, A.; Karaś, M.; Rybczyńska, K. Stevia Rebaudiana Bert. Leaf Extracts as a Multifunctional Source of Natural Antioxidants. *Molecules* **2015**, *20*, 5468–5486. [CrossRef]
51. Callow, E.H. Comparative studies of meat. II. The changes in the carcass during growth and fattening, and their relation to the chemical composition of the fatty and muscular tissues. *J. Agric. Sci.* **1948**, *38*, 174–199. [CrossRef]
52. Mehdi, Y.; Clinquart, A.; Hornick, J.-L.; Cabaraux, J.-F.; Istasse, L.; Dufrasne, I. Meat composition and quality of young growing Belgian Blue bulls offered a fattening diet with selenium enriched cereals. *Can. J. Anim. Sci.* **2015**, *95*, 465–473. [CrossRef]
53. Waldenstedt, L. Nutritional factors of importance for optimal leg health in broilers: A review. *Anim. Feed Sci. Technol.* **2006**, *126*, 291–307. [CrossRef]
54. Zhang, S.; Xie, Y.; Li, M.; Yang, H.; Li, S.; Li, J.; Xu, Q.; Yang, W.; Jiang, S. Effects of Different Selenium Sources on Meat Quality and Shelf Life of Fattening Pigs. *Animals* **2020**, *10*, 615. [CrossRef]
55. Choi, J.S.; Lee, J.H.; Lee, H.J.; Jang, S.S.; Lee, J.J.; Choi, Y.I. Effect of stevia and charcoal as an alternative to antibiotics on carcass characteristics and meat quality in finishing pigs. *Food Sci. Anim. Resour.* **2012**, *32*, 835–841. [CrossRef]
56. Smith, G.C.; Belk, K.E.; Sofos, J.N.; Tatum, J.D.; Williams, S.N. Economic implications of improved color stability in beef. In *Antioxidants in Muscle Foods: Nutritional Strategies to Improve Quality*; Decker, E.A., Faustman, C., Lopez-Bote, C.J., Eds.; Wiley Interscience: New York, NY, USA, 2000; pp. 397–426.
57. Macit, M.; Aksakal, V.; Emsen, E.; Aksu, M.I.; Karaoglu, M.; Esenbuga, N. Effects of vitamin E supplementation on performance and meat quality traits of Morkaraman male lambs. *Meat Sci.* **2003**, *63*, 51–55. [CrossRef]
58. Boiago, M.M.; Borba, H.; Leonel, F.R.; Giampietro-Ganeco, A.; Ferrari, F.B.; Stefani, L.M.; Souza, P.A. de Sources and levels of selenium on breast meat quality of broilers. *Cienc. Rural* **2014**, *44*, 1692–1698. [CrossRef]
59. Boonkaewwan, C.; Burodom, A. Anti-inflammatory and immunomodulatory activities of stevioside and steviol on colonic epithelial cells: Anti-inflammatory and immunomodulatory activities of stevia compounds. *J. Sci. Food Agric.* **2013**, *93*, 3820–3825. [CrossRef] [PubMed]
60. Chatsudthipong, V.; Muanprasat, C. Stevioside and related compounds: Therapeutic benefits beyond sweetness. *Pharmacol. Ther.* **2009**, *121*, 41–54. [CrossRef] [PubMed]
61. Ruiz, J.C.R.; Ordoñez, Y.B.M.; Basto, Á.M.; Campos, M.R.S. Antioxidant capacity of leaf extracts from two *Stevia rebaudiana* Bertoni varieties adapted to cultivation in Mexico. *Nutr. Hosp.* **2015**, *31*, 1163–1170. [CrossRef]
62. Zheng, G.; Xu, L.; Wu, P.; Xie, H.; Jiang, Y.; Chen, F.; Wei, X. Polyphenols from longan seeds and their radical-scavenging activity. *Food Chem.* **2009**, *116*, 433–436. [CrossRef]
63. Mukai, K.; Nagai, S.; Ohara, K. Kinetic study of the quenching reaction of singlet oxygen by tea catechins in ethanol solution. *Free Radic. Biol. Med.* **2005**, *39*, 752–761. [CrossRef]
64. Brigelius-Flohé, R.; Maiorino, M. Glutathione peroxidases. *Biochim. Biophys. Acta* **2013**, *1830*, 3289–3303. [CrossRef]
65. Berchielli, T.T.; Vaz Pires, A.; Oliveira, S.G. Nutrição de Ruminantes. *Jaboticabal Funep* **2006**, *1*, 200–583.
66. Richard, D.; Bausero, P.; Schneider, C.; Visioli, F. Polyunsaturated fatty acids and cardiovascular disease. *Cell. Mol. Life Sci.* **2009**, *66*, 3277–3288. [CrossRef] [PubMed]
67. Netto, A.S.; Zanetti, M.A.; Claro, G.R.D.; de Melo, M.P.; Vilela, F.G.; Correa, L.B. Effects of Copper and Selenium Supplementation on Performance and Lipid Metabolism in Confined Brangus Bulls. *Asian Australas. J. Anim. Sci.* **2014**, *27*, 488–494. [CrossRef] [PubMed]
68. Lourenço, M.; Cardozo, P.W.; Calsamiglia, S.; Fievez, V. Effects of saponins, quercetin, eugenol, and cinnamaldehyde on fatty acid biohydrogenation of forage polyunsaturated fatty acids in dual-flow continuous culture fermenters1. *J. Anim. Sci.* **2008**, *86*, 3045–3053. [CrossRef]
69. Li, Z.T.; Lin, T.; Shen, J.X.; Chen, W.Y.; Sun, C. Effect of rosemary on antibacterial and freshness of cold meat. *Food Res. Dev.* **2017**, *38*, 181–186.
70. Zhou, G.H.; Xu, X.L.; Liu, Y. Preservation technologies for fresh meat—A review. *Meat Sci.* **2010**, *86*, 119–128. [CrossRef]

71. Kostadinović, L.; Lević, J.; Popović, S.T.; Čabarkapa, I.; Puvača, N.; Djuragic, O.; Kormanjoš, S. Dietary inclusion of Artemisia absinthium for management of growth performance, antioxidative status and quality of chicken meat. *Eur. Poult. Sci.* **2015**, *79*. [CrossRef]
72. Hęś, M. Protein-lipid interactions in different meat systems in the presence of natural antioxidants—A review. *Polish J. Food Nutr. Sci.* **2017**, *67*, 5–18. [CrossRef]
73. Mikulski, D.; Jankowski, J.; Zdun Czyk, Z.; Wróblewska, M.; Sartowska, K.; Majewska, T. The effect of selenium source on performance, carcass traits, oxidative status of the organism, and meat quality of turkeys. *J. Anim. Feed Sci.* **2009**, *18*, 518–530. [CrossRef]

MDPI
St. Alban-Anlage 66
4052 Basel
Switzerland
Tel. +41 61 683 77 34
Fax +41 61 302 89 18
www.mdpi.com

Foods Editorial Office
E-mail: foods@mdpi.com
www.mdpi.com/journal/foods

www.ingramcontent.com/pod-product-compliance
Lightning Source LLC
LaVergne TN
LVHW070547100526
838202LV00012B/408